Annals of Mathematics Studies

Number 59

ANNALS OF MATHEMATICS STUDIES

Edited by Robert C. Gunning, John C. Moore, and Marston Morse

1. Algebraic Theory of Numbers, *by* HERMANN WEYL
3. Consistency of the Continuum Hypothesis, *by* KURT GÖDEL
11. Introduction to Nonlinear Mechanics, *by* N. KRYLOFF *and* N. BOGOLIUBOFF
20. Contributions to the Theory of Nonlinear Oscillations, Vol. I, *edited by* S. LEFSCHETZ
21. Functional Operators, Vol. 1, *by* JOHN VON NEUMANN
24. Contributions to the Theory of Games, Vol. I, *edited by* H. W. KUHN *and* A. W. TUCKER
25. Contributions to Fourier Analysis, *edited by* A. ZYGMUND, W. TRANSUE, M. MORSE, A. P. CALDERON, *and* S. BOCHNER
28. Contributions to the Theory of Games, Vol. II, *edited by* H. W. KUHN *and* A. W. TUCKER
30. Contributions to the Theory of Riemann Surfaces, *edited by* L. AHLFORS *et al.*
33. Contributions to the Theory of Partial Differential Equations, *edited by* L. BERS, S. BOCHNER, *and* F. JOHN
34. Automata Studies, *edited by* C. E. SHANNON *and* J. MCCARTHY
38. Linear Inequalities and Related Systems, *edited by* H. W. KUHN *and* A. W. TUCKER
39. Contributions to the Theory of Games, Vol. III, *edited by* M. DRESHER, A. W. TUCKER *and* P. WOLFE
40. Contributions to the Theory of Games, Vol. IV, *edited by* R. DUNCAN LUCE *and* A. W. TUCKER
41. Contributions to the Theory of Nonlinear Oscillations, Vol. IV, *edited by* S. LEFSCHETZ
42. Lectures on Fourier Integrals, *by* S. BOCHNER
43. Ramification Theoretic Methods in Algebraic Geometry, *by* S. ABHYANKAR
44. Stationary Processes and Prediction Theory, *by* H. FURSTENBERG
45. Contributions to the Theory of Nonlinear Oscillations, Vol. V, *edited by* L. CESARI, J. LASALLE, *and* S. LEFSCHETZ
46. Seminar on Transformation Groups, *by* A. BOREL *et al.*
47. Theory of Formal Systems, *by* R. SMULLYAN
48. Lectures on Modular Forms, *by* R. C. GUNNING
49. Composition Methods in Homotopy Groups of Spheres, *by* H. TODA
50. Cohomology Operations, *lectures by* N. E. STEENROD, *written and revised by* D. B. A. EPSTEIN
51. Morse Theory, *by* J. W. MILNOR
52. Advances in Game Theory, *edited by* M. DRESHER, L. SHAPLEY, *and* A. W. TUCKER
53. Flows on Homogeneous Spaces, *by* L. AUSLANDER, L. GREEN, F. HAHN, *et al.*
54. Elementary Differential Topology, *by* J. R. MUNKRES
55. Degrees of Unsolvability, *by* G. E. SACKS
56. Knot Groups, *by* L. P. NEUWIRTH
57. Seminar on the Atiyah-Singer Index Theorem, *by* R. S. PALAIS
58. Continuous Model Theory, *by* C. C. CHANG *and* H. J. KEISLER
59. Lectures on Curves on an Algebraic Surface, *by* DAVID MUMFORD
60. Topology Seminar, Wisconsin, 1965, *edited by* R. H. BING *and* R. J. BEAN

LECTURES ON CURVES ON AN ALGEBRAIC SURFACE

BY

David Mumford

WITH A SECTION BY

G. M. Bergman

PRINCETON, NEW JERSEY

PRINCETON UNIVERSITY PRESS

1966

L. C. Card: 66-17705

Printed in the United States of America

DEDICATION

The contributors to this volume dedicate their work to the memory of

M. K. Fort, Jr.

whose warmth and good will have been felt by the entire mathematical community.

INTRODUCTION

These notes are being printed in exactly the form in which they were first written and distributed: as class notes, supplementing and working out my oral lectures. As such, they are far from polished and ask a lot of the reader. In the words of the ex-editor of a well-known journal they are written in a style "seldom seen except in personal letters between close friends." Be that as it may, my hope is that a well-intentioned reader will still be able to penetrate these notes and learn something of the beautiful geometry on an algebraic surface.

It was expected, when these notes were written, that the reader had the following background: he had taken a graduate course in commutative algebra, he had studied some Algebraic Geometry and, in particular, he had some acquaintance with the theory of curves, and the theory of schemes, and of their cohomology (e.g., Dieudonne's Maryland and Montreal Lecture Notes). Nonetheless, both to fix ideas, and to prove some specialized results that are needed later, Lectures 3-10 are devoted to a quick and rather breezy digression into the general theory of schemes. Lecture 11 summarizes what we need from the theory of curves. I apologize to any reader who, hoping that he would find here in these 60 odd pages an easy and concise introduction to schemes, instead became hopelessly lost in a maze of unproven assertions and undeveloped suggestions. From Lecture 12 on, we have proven everything that we need.

The goal of these lectures is a complete clarification of one "theorem" on Algebraic surfaces: the so-called completeness of the characteristic linear system of a good complete algebraic system of curves, on a surface F. If the characteristic is 0, this theorem was first proven by Poincaré (cf. References) in 1910 by analytic methods. Until about 1960, no algebraic proof of this purely algebraic theorem was known.* In 1955, Igusa had shown that the theorem, as stated, was false in characteristic p thus making the theorem appear even more analytic in nature. But about 1960, a truly amazing development occurred: in the course of working out the master plan that he had laid out for Algebraic Geometry—incorporating some of the key ideas of Kodaira's and Spencer's deformation theory—Grothendieck had occasion to write out some of the Corollaries of his theory (cf. his Bourbaki exposé 221, pp. 23-24). Putting his results together with a

* Although an endless and depressing controversy obscured this fact.

result of Cartier—that group schemes in characteristic 0 are reduced—one finds that this old problem has been completely solved: a) a purely algebraic proof is available in characteristic 0, b) all the machinery is ready at hand for obtaining, in characteristic p, necessary and sufficient conditions for the validity of the theorem. What was the key, the essential point which the Italians had overlooked? There is no doubt at all that it is the systematic use of nilpotent elements: in particular, a systematic analysis of all families of curves on a surface over a parameter space with only one point, but with non-trivial nilpotent structure sheaf. The Italians had, in a sense, done this, but only when the ring of functions on the base was Study's ring of dual numbers $k[\varepsilon]/(\varepsilon^2)$. This is the same as looking at first-order deformations of a curve. But they ignored higher order nilpotents and higher order deformations.

The outline of these lectures is as follows—Lectures 1 and 2 give an intuitive introduction to the problem and present in outline 2 analytic proofs. Lectures 3 through 10 recall basic notions about schemes. Lectures 11 through 21 deal with basic questions on the theory of surfaces. In particular, they give a construction of universal families of curves on a surface—the so-called Hilbert scheme; and of universal families of divisor classes on a surface—the so-called Picard scheme. Lectures 22 through 27 deal with the application of the whole theory to the main problem: these include a long lecture by G. Bergman giving a self-contained description of the Witt ring schemes.

I would like to call attention to several generalizations and applications of our results which were omitted so as to get directly to the main result.

a) The method by which we have constructed the universal family of curves on a surface F gives without any change a construction of the universal flat family of subschemes of any scheme X, projective over a noetherian S, i.e., of the Hilbert scheme. In particular, the explicit estimates obtained in Lecture 14 enable ont to carry through this construction—which is Grothendieck's original construction—without the indirect arguments using the concept of "limited families" which he used (cf. his "Fondements").

b) The method by which we have constructed the Picard scheme of a surface F generalizes so as to construct the Picard scheme of any scheme X, projective and flat over a noetherian S, whose geometric fibres over S are reduced and connected and such that the components of its actual fibres over S are absolutely irreducible. This construction is related to the one I outlined at the International Congress of 1962, and ties up with the methods used in Chapters 3 and 7 of my book *Geometric Invariant Theory*.

c) One can use the results of Lecture 18 to give a very easy proof of the Riemann Hypothesis for curves over finite fields. This is the proof of Mattuck-Tate (cf. References). If you have read through Lecture 18, and know the formulation of the Riemann Hypothesis via the Frobenius morphism, you can read their paper without difficulty and you should.

Cambridge
March, 1966

CONTENTS

LECTURE 1

RAW MATERIAL ON CURVES ON SURFACES, AND THE PROBLEMS SUGGESTED

We shall be concerned entirely with algebraic geometry over a fixed algebraically closed field k (of arbitrary characteristic). Our chief purpose is to study the geometry on a non-singular algebraic surface F, projective over k, and, in particular, the families of curves C on F.

By a curve we mean either a finite sum of irreducible, 1-dimensional subvarieties of F, with positive multiplicity: $\Sigma\, n_i C_i$, or a sheaf of principal ideals on F. [These are equivalent concepts—for precise definitions, cf. Lecture 9.]

Example 1: $F = \mathbf{P}_2$. Then, as is well-known, every curve C on $\mathbf{P}_2$ is defined by a homogeneous form $F(x_0, x_1, x_2)$. In particular, one can attach to C its degree d, i.e., the degree of F, and the family of all curves of degree d is parametrized by the set of all F of degree d, up to scalars: i.e., by a projective space of dimension

$$\frac{(d+1)(d+2)}{2} - 1$$

Example 2: $F = \mathbf{P}_1 \times \mathbf{P}_1$ (i.e., a quadric in $\mathbf{P}_3$). Then every curve C on F is defined by a bi-homogeneous form

$$F(x_0, x_1; y_0, y_1)$$

with two degrees d and e. d and e can be interpreted as the degrees of the coverings

$$p_1, p_2\colon\ C \to \mathbf{P}_1$$

given by the two projections of $\mathbf{P}_1 \times \mathbf{P}_1$ onto $\mathbf{P}_1$. Again, for every d and e, there is a single family of curves parametrized by a projective space, this time of dimension;

$$(d+1)(e+1) - 1$$

The phenomenon of the last two examples can be generalized by the concept of a linear system. If f is an algebraic function on F, let, as usual, (f) stand for the formal sum:

$$\sum_{\substack{\text{all 1-dimensional} \\ \text{irreducible subvarieties} \\ E}} \operatorname{ord}_E(f) \cdot E$$

where $\operatorname{ord}_E(f)$ is the order of the zero or pole of f at E. Then associated to any curve C one has the vector space of functions with poles only at C:

$$\mathfrak{L}(C) = \{f \mid (f) + C \geq 0\}$$

(Here $\Sigma n_i E_i \geq 0$ means all $n_i \geq 0$.) If $f_0, \ldots, f_n$ are a basis of $\mathfrak{L}(C)$, one then can define the following family of curves, which contains C:

$$C_\alpha = (\Sigma \alpha_i f_i) + C$$

Since C_α only depends on the ratios of the α_i, this is an irreducible family of curves parametrized by a projective space of dimension:

$$\dim \mathfrak{L}(C) - 1$$

Linear systems are the simplest families of curves on a surface F and the only type occurring in Examples 1 and 2.

<u>Definition:</u> Two curves C_1 and C_2 are <u>linearly equivalent</u> if equivalently:

i) $\exists$ a function f on F such that $(f) = C_1 - C_2$, or

ii) C_1, C_2 are in the same linear system.

We write $C_1 \equiv C_2$ for this concept.

<u>Example 3:</u> Let $\mathcal{E}$ be an elliptic curve (over k), and let $F = \mathbf{P}_1 \times \mathcal{E}$. Again, given a curve C on F, we can assign to C two degrees d and e, as the orders of the coverings

$$C \to \mathbf{P}_1; \quad C \to \mathcal{E}$$

obtained by projecting. Both $d \geq 0$ and $e \geq 0$ and either $d > 0$ or $e > 0$.

<u>Case i)</u> $d = 0$: Then C is of the form $\Sigma_{i=1}^{e} P_i \times \mathcal{E}$, and all these C form a single e-dimensional linear system.

<u>Case ii)</u> $d > 0$: The set of all C of type (d, e) forms an irreducible $d(e + 1)$-dimensional family of curves, but it is not a linear system. Rather it is fibred by $d(e + 1)-1$-dimensional linear subfamilies.

<u>Definition:</u> Two curves C_1, C_2 are <u>algebraically equivalent</u> if C_1 and C_2 are both contained in one family of curves parametrized by a connected variety.

With this terminology, we can say that on $\mathbf{P}_1 \times \mathcal{E}$, algebraic and linear equivalence differ. Another point to notice is that the dimension formula in Case ii) does not specialize to the dimensional formula in Case i) when $d = 0$: this is the phenomenon of <u>superabundance</u>.

Example 4: Let γ be a "generic" curve of genus 2, i.e., a double covering of $\mathbf{P}_1$ branched at six points with independent transcendental coordinates over the prime field (if char. $\neq 2$). Let F be the jacobian of γ. Recall that

(1) F is a non-singular algebraic surface,

(2) F is also an algebraic group,

(3) in a natural way, γ itself is a curve on F.

It turns out that every curve C on F is algebraically equivalent to a curve $d\gamma$, for a suitable positive integer d. Moreover, C is linearly equivalent to a suitable translation of $d\gamma$ (in the sense of the given group structure). The set of all curves algebraically equivalent to $d\gamma$ is an irreducible family of dimension $d^2 + 1$, and its linear sub-families have dimension $d^2 - 1$. In fact, one can define a map:

$$F \rightarrow \left[\frac{\text{all curves alg. equivalent to } d\gamma}{\text{linear equivalence}} \right]$$

where a $\mapsto$ image of $d\gamma$ under translation by a. In fact, this map factors as follows:

$$F \xrightarrow{\text{mult. by } d} F \xrightarrow{\text{bijection}} \left[\frac{\text{curves alg. equivalent to } d\gamma}{\text{linear equivalence}} \right]$$

This indicates a general point: the set [algebraic equivalence modulo linear equivalence], tends to be independent of the family of curves considered.

One should contrast this surface F with its "Kummer" counterpart K: this is defined as the double covering of $\mathbf{P}_2$ branched in a generic sextic curve (char. $\neq 2$). Here all curves are <u>linearly</u> equivalent to $d \cdot h$, where h is the inverse image of a line in $\mathbf{P}_2$, and the dimension of this family is $d^2 + 1$ (as above). It is similar to F also in that (a) $(\gamma^2) = 2$ on F, $(h^2) = 2$ on K [(D^2) denotes self-intersection—cf. Lecture 12], and (b) both F and K admit double differentials with neither zeros nor poles. This K is of the same type as the <u>quartic</u> surfaces in $\mathbf{P}_3$.

In fact, we have touched briefly on every class of algebraic surfaces admitting a double differential <u>with no zeros</u> (i.e., an anti-canonical linear system): for reasons stemming from Serre duality, the geometry on these surfaces is particularly simple. To bring out some further features of surfaces, we will discuss another rational surface:

Example 5: Let F be the surface obtained by blowing up two points P_1, P_2 in $\mathbf{P}_2$ [or by blowing up one point in $\mathbf{P}_1 \times \mathbf{P}_1$]. Let E_1 and E_2 be the rational curves which are the inverse images of P_1 and P_2 on F. Let ℓ be the line in $\mathbf{P}_2$ from P_1 to P_2, and let D be the curve on F which is the closure of the inverse image of $\ell - P_1 - P_2$. Then to every curve C on F, one can attach <u>three</u> characters k_1, k_2, and ℓ, where k_1, k_2 and ℓ are non-negative and not all zero; and the set of all curves with characters k_1, k_2, ℓ form the single linear system containing

$$k_1E_1 + k_2E_2 + \ell D$$

But unlike the situation on $\mathbf{P}_1 \times \mathbf{P}_1$, not all these systems are "good" systems of curves.

Case i) If $\ell \geq k_1$, $\ell \geq k_2$ and $k_1 + k_2 \geq \ell$, then none of the three curves E_1, E_2, or D is a component of all curves in the linear system containing $k_1E_1 + k_2E_2 + \ell D$, and this linear system has the predictable dimension:

$$(*) \qquad \frac{(\ell+1)(\ell+2)}{2} - \frac{(\ell-k_1)(\ell-k_1+1)}{2} - \frac{(\ell-k_2)(\ell-k_2+1)}{2} - 1$$

Case ii) If $\ell < k_1$, $\ell < k_2$, or $k_1 + k_2 < \ell$, then one of the three curves E_1, E_2, or D is a component of all the curves in question, and, in general, this family is also superabundant, i.e., its dimension is bigger than that predicted by (*).

Another way of telling the "good" from the "bad" systems of curves is this:

$$\left\{\begin{array}{l}\text{the system of curves}\\ \text{linearly equivalent}\\ \text{to } k_1E_1 + k_2E_2 + \ell D\\ \text{is the family of hyper-}\\ \text{plane sections of } F\\ \text{for some embedding of } F\\ \text{in } \mathbf{P}_N\end{array}\right\} \iff \begin{array}{l}\ell > k_1\\ \ell > k_2\\ k_1 + k_2 > \ell\end{array}$$

Here the condition on the left defines the notion: $k_1E_1 + k_2E_2 + \ell D$ is very ample.

With all this data before us, what questions emerge as the natural ones to pose in studying the curves on a general surface F? I think four basic lines of study are suggested:

(i) the problem of Riemann-Roch: Given a curve C, to determine the dimension of the linear system of curves containing C. We shall see below that this is equivalent to the problem of computing

$$\dim H^0(\mathfrak{L})$$

where $\mathfrak{L}$ is a sheaf on F, locally isomorphic to the sheaf $\underline{o}_F$ of regular functions.

(ii) the problem of Picard: To describe the family of all algebraic deformations of a curve C modulo its linear subfamilies. It turns out that this quotient is independent of C, if C is good, and this quotient leads to the Picard scheme and/or variety.

(iii) <u>Good vs. Bad curves:</u> What makes C good and bad? One can ask when is C very ample, when is C superabundant, what are the really bad "exceptional" C which play the role of E_1, E_2 and D in Example 5 above? Particularly significant is the question of the "regularity of the adjoint" (= "Kodaira's vanishing theorem") cf. Lecture 14.

(iv) <u>the components of the set of all curves C on F</u>: Especially, what finiteness statements can be made? Examples are the theorem of the base of Neron and Severi, and the theorem that only a finite number of components represent curves of any given degree.

LECTURE 2

THE FUNDAMENTAL EXISTENCE PROBLEM AND

TWO ANALYTIC PROOFS

We shall analyze problem ii) more closely. The real nature of the problem becomes clearer when one passes from curves to divisors. By a divisor on F we mean either a finite sum of irreducible, 1-dimensional subvarieties, with (positive or negative) multiplicity: $\Sigma n_i C_i$, $n_i \in \mathbf{Z}$, or a sheaf of fractional ideals, i.e., a coherent subsheaf of the constant sheaf $\underline{K}$:

$$K(U) = \text{function field } k(F), \text{ all } U$$

(cf. Lecture 9 for precise definitions). The set of all divisors on F forms a group, which we denote $G(F)$. Put:

$G_a(F)$ = subgroup of divisors of the form $C_1 - C_2$, where C_1, C_2 are algebraically equivalent curves,

$G_\ell(F)$ = subgroup of divisors of the form $C_1 - C_2$, where $C_1 \equiv C_2$, or, equivalently, the subgroup of divisors of form (f), $f \in k(F)$.

Now if C is any curve on F, and $\{C_\alpha | \alpha \in S\}$ is the family of all curves algebraically equivalent to $C = C_0$, one can define a map:

$$S/\begin{matrix}\text{modulo linear}\\ \text{subfamilies}\end{matrix} \longrightarrow G_a(F)/G_\ell(F)$$

by mapping α to the divisor $C_\alpha - C_0$. One checks immediately that it is always injective, and it can be shown that for sufficiently "good" (?!) curves, it is surjective. For this reason, problem (ii) becomes independent of C, in most cases, and asks simply—what is the structure and dimension of the group $G_a(F)/G_\ell(F)$ invariantly attached to F ?

Again without proofs, we would like to mention the cohomological interpretation of these groups:

Let $\underline{o}^*$ = sheaf of units in the structure sheaf $\underline{o}$
$\underline{K}^*$ = sheaf of units in $\underline{K}$.

Then:

$$0 \to \underline{o}^* \to \underline{K}^* \to \underline{K}^*/\underline{o}^* \to 0$$

leads to:

$$\begin{array}{ccccccc} 0 \to H^0(\underline{K}^*)/k^* & \to & H^0(\underline{K}^*/\underline{o}^*) & \to & H^1(\underline{o}^*) & \to & 0 \\ \wr\| & & \wr\| & & & & \\ G_\ell(F) & & G(F) & & & & \end{array}$$

Therefore, $G_a(F)/G_\ell(F)$ is a subgroup of $H^1(\underline{o}^*)$, the so-called <u>Picard group</u> of F (definable on any ringed-space).

Now the work of Castelnuovo and Matsusaka has shown that the group $G_a(F)/G_\ell(F)$ can be given, in a natural way, the structure of an algebraic group—in fact, an abelian variety. The essential point is, however, what is the dimension? Here we have an existence problem: can we predict the dimension of the set of solutions of an essentially non-linear problem by means of some linear data, e.g., the cohomology of a coherent sheaf? It was conjectured by Severi that:

(A) $$\dim G_a(F)/G_\ell(F) = \dim H^1(\underline{o})$$

where $\underline{o}$ = structure sheaf on F, (in his language, $q = p_g - p_a$). This was proven by Poincaré in 1909, when $k = \mathbf{C}$, and was disproven by Igusa in 1953, when $\mathrm{char}(k) \neq 0$.

The simplest way to motivate (A) is to note that the term on the left is a subgroup of $H^1(\underline{o}^*)$, and to guess that there should be some kind of "exponential" from $H^1(\underline{o})$ to $H^1(\underline{o}^*)$, (cf. below). A second way is to transform (A) into a statement concerning the deformations of a curve C on F, and in this form, it is a special case of the general Kodaira-Spencer existence problem for deformations. To see this, suppose again that $\{C_\alpha \mid \alpha \in S\}$ is a family of deformations of $C = C_0$. Let N be the sheaf of sections of the normal bundle to C in F (assume C is non-singular). Then there is a fundamental characteristic map:

$$\left\{ \begin{array}{l} \text{Tangent Space} \\ \text{to } S \text{ at } \alpha = 0 \end{array} \right\} \xrightarrow{\rho} H^0(N) \quad .$$

Roughly speaking, a small neighborhood of C in F is nearly isomorphic to the normal bundle to C in F, while a curve C_α, for α near 0, defines a section of this neighborhood: as $\alpha \to 0$ therefore, these curves can be asymptotically identified with section of the normal bundle to C in F. The key existence problem is now:

(B) for suitable $\{C_\alpha\}$, ρ is bijective

Incidentally, in this form, the conjecture can be equally well posed for subvarieties in other varieties of arbitrary codimension, e.g., for deformations of curves in $\mathbf{P}_3$. Unfortunately, it is false even in char. 0 for some pathological space curves.

To connect conjectures (A) and (B), we use the exact sequence of sheaves:

$$0 \to \underline{o} \to \underline{o}(C) \xrightarrow{\varphi} N \to 0$$

where

$$\left\{ \begin{array}{l} \underline{o}(C) = \text{sheaf of functions with simple poles at } C \\ \varphi \text{ maps the function } A/f \text{ into the normal vector} \\ \quad \text{field } X \text{ such that } X(df) = A \end{array} \right.$$

(Here $f = 0$ is a local equation of C.)

Then one can show that, for "good" curves C, there is a commutative diagram:

$$\begin{array}{ccccccccc} 0 & \longrightarrow & H^0(\underline{o}(C))/k & \longrightarrow & H^0(N) & \longrightarrow & H^1(\underline{o}) & \longrightarrow & 0 \\ & & \sigma \uparrow & & \rho \uparrow & & \uparrow \tau & & \\ 0 & \to & \left\{ \begin{array}{l} \text{tang. sp. to} \\ S_0 \text{ at } \alpha = 0 \end{array} \right\} & \longrightarrow & \left\{ \begin{array}{l} \text{tang. sp. to} \\ S \text{ at } \alpha = 0 \end{array} \right\} & \to & \left\{ \begin{array}{l} \text{tang. sp. to} \\ G_a(F)/G_\ell(F) \\ \text{at } 0 \end{array} \right\} & \to & 0 \end{array}$$

where $S_0 \subset S$ is the linear subfamily through 0, and S mod linear equivalence is identified to $G_a(G)/G_\ell(F)$. Moreover, σ is <u>always</u> an isomorphism. Therefore ρ is bijective if and only if τ is bijective, and (A) is equivalent to (B).

Before passing to our systematic discussion, I would like to sketch two proofs of conjecture (A) in case $k = \mathbf{C}$.

<u>Proof I</u> (GAGA): Let $\underline{o}_h$ = sheaf of holomorphic functions on F, and let $\underline{o}_h^* \subset \underline{o}_h$ be the subsheaf of units. Then the exponential defines an exact sequence:

$$0 \longrightarrow \mathbf{Z} \longrightarrow \underline{o}_h \xrightarrow{e^{2\pi i(\ )}} \underline{o}_h^* \longrightarrow 0 \ .$$

hence:

$$H^1(\underline{o}_h) \xrightarrow{\exp} H^1(\underline{o}_h^*) \ .$$

But by GAGA:

$$H^1(\underline{o}) \xrightarrow{\sim} H^1(\underline{o}_h)$$
$$H^1(\underline{o}^*) \xrightarrow{\sim} H^1(\underline{o}_h^*) \ ,$$

hence there is an induced exponential on the algebraic level from $H^1(\underline{o})$ to $H^1(\underline{o}^*)$.

<u>Proof II</u> (POINCARÉ): In this proof, the only GAGA-type assertion we require will be that meromorphic functions defined on the whole of the complex projective line $\mathbf{P}_1$ are algebraic.

We pick a nice pencil of curves C_t on F, $t \in \mathbf{P}_1$. Let J_t be the jacobian (or generalized jacobian) of C_t, and let $J = \bigcup J_t$ be the variety of <u>all</u> the J_t's. Let p = genus (C_t), and $q = \dim H^1(\underline{o})$. If we define a q-dimensional family of sections of J over $\mathbf{P}_1$, then we can define, for each section s, a 0-cycle $\mathfrak{A}_t(s)$ of degree p on each C_t,

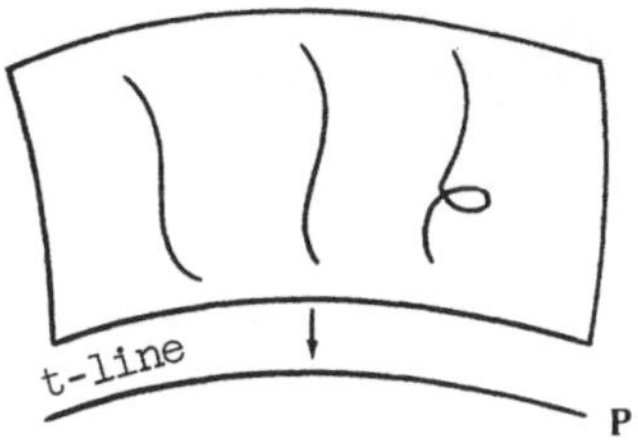

hence, a curve $D(s)$ on F such that $D(s) \cdot C_t = \mathfrak{A}_t(s)$. One can prove that this gives a q-dimensional family of non-linearly equivalent divisors. Moreover, by our remark above, it is the same to construct these sections algebraically or holomorphically.

Recall that J_t is obtained by considering the integrals of the simple differentials with no poles on C_t, modulo their periods: or, what is the same,

$$J_t \cong \frac{\text{Dual space of } H^0(\Omega^1_{C_t})}{\text{Linear functionals given by periods}}$$

where $\Omega^1_{C_t}$ = sheaf of simple differentials on C_t, with no poles. By Serre duality on C_t,

$$\text{Dual space of } H^0(\Omega^1_{C_t}) \cong H^1(\underline{o}_{C_t}) \quad .$$

Therefore, one obtains the q-dimensional family simply by choosing $\alpha \in H^1(\underline{o})$, restricting α to $H^1(\underline{o}_{C_t})$, for every t, and mapping this element to a point of J_t by the above identifications.

LECTURE 3

PRE-SCHEMES AND THEIR ASSOCIATED "FUNCTOR OF POINTS."

We first recall the most basic definitions and results in the theory of pre-schemes:

1° Pre-schemes are (like all structured geometric objects) topological spaces X, endowed with a sheaf of rings $\underline{o}_X$ (or $\underline{o}$), whose stalks are local rings. Their characteristic property is that they admit an open covering $\{U_i\}$ such that $(U_i, \underline{o}|U_i)$ is isomorphic (for all i) to one of the standard pre-schemes:

$X = \mathrm{Spec}(A) =$
- a) as point set, the set of prime ideals $p \subset A$
- b) as topological spaces, a basis of open sets is given by the subsets $X_f = \{p \mid f \notin p\}$, for $f \in A$
- c) its structure sheaf is defined by: $\Gamma(X_f, \underline{o}_X) = A_{(f)}$

for *any* commutative ring A with 1.

Pre-schemes, as local ringed spaces, are very "un-classical" in appearance. In the first place, they are full of non-closed points: we shall say that a closed subset Z in a pre-scheme X is irreducible if it is not the union of two properly smaller closed subsets. Then one finds: given any irreducible closed subset Z in X, there is one and only one point $z \in Z$ such that Z is the closure of z. This z is called the *generic point* of Z. Since, if A is a noetherian ring, the closed subsets of Spec (A) will satisfy the descending chain condition, a scheme such as Spec (A) has in general plenty of irreducible closed subsets, hence plenty of non-closed points. In case all the local rings $\underline{o}_x$ are noetherian, one can introduce a numerical measure of the non-closedness of points by:

$$\mathrm{codim}\,(x) = \text{Krull dimension of } \underline{o}_x,$$

and, consequently also for the size of irreducible subsets:

$$\mathrm{codim}\,(Z) = \text{codimension of the generic point of } Z.$$

This has the good property: if $Z_1 \subsetneq Z_2$ are two closed irreducible subsets with generic points z_1, z_2 (i.e., z_1 in the closure of z_2, but not vice versa) then:

$$\left.\begin{array}{l}\text{codim } Z_1 > \text{codim } Z_2 \\ \text{codim } z_1 > \text{codim } z_2\end{array}\right\} \text{ and}$$

[For proof, check that $\underline{o}_{z_2}$ is a localization $(\underline{o}_{z_1})_p$ for a prime ideal $p \subset \underline{o}_{z_1}$, p not maximal.]

The following simple property, given directly in terms of the data of a local ringed space, distinguishes pre-schemes from most other local ringed spaces:

Proposition 1: Let X be a local ringed space, $x \in X$, and $\underline{o}_x$ the local ring at x. Let

$$S_x = \{y \in X \mid x \text{ is in the closure of } y\}.$$

Then if X is a pre-scheme, S_x with its induced topology and sheaf of rings is isomorphic to Spec $(\underline{o}_x)$.

[Proof: Reduce to the case X affine, where it is clear.]

Even leaving out non-closed points, pre-schemes are very un-Hausdorff: Look at $X = \text{Spec } k[X]$, the affine line over an algebraically closed field k. The prime ideal (0) gives the generic point, and, for all $\alpha \in k$, the prime ideal $(X - \alpha)$ gives a closed point $P_\alpha \in X$. These are the only points of X, and every open set is of the form:

$$X - \bigcup_{\alpha \in \text{(finite) set}} P_\alpha .$$

In particular, no two open sets are disjoint.

Another unclassical aspect of the pre-schemes should be stressed at the outset. Just as in any local ringed space X, a section $f \in \Gamma(U, \underline{o}_X)$, for $U \subset X$ open, can be regarded as a function on U. At a point $x \in U$, its values are taken in the residue field $\mathfrak{K}(x)$ of the stalk $\underline{o}_x$, and the value of f is:

$$f(x) = \text{image of } f \text{ in } \mathfrak{K}(x) .$$

However, it is quite possible that $f \neq 0$ while $f(x) = 0$ for all x. It is this aspect of pre-schemes which was most scandalous when Grothendieck defined them. Suppose $U = \text{Spec }(A)$, and $f \in A$. Then, in fact, such sections f are easy to describe:

The following are equivalent:

i) $f(x) = 0$, all $x \in U$

ii) $f \in p$, all prime ideals $p \subset A$,

iii) f is nilpotent, (since in A, $\sqrt{(0)} = \bigcap_{\text{all prime ideals}} p$!) .

2° If X and Y are pre-schemes, the morphisms f from X to Y are taken to be arbitrary morphisms of X and Y as <u>local ringed spaces</u>; i.e., continuous maps

$$f': X \to Y$$

plus homomorphisms

$$f'': \underline{o}_Y \to f'_*(\underline{o}_X)$$

inducing local homomorphisms on the stalks. The key result in interpreting these morphisms concretely is:

THEOREM 1: Let X be any pre-scheme, and let Y = Spec (A). To any morphism f: $X \to Y$, one can attach a homomorphism:

$$A \xrightarrow{\sim} \Gamma(Y, \underline{o}_Y) \longrightarrow \Gamma(Y, f_*\underline{o}_X) \xrightarrow{\sim} \Gamma(X, \underline{o}_X) .$$

This sets up an isomorphism

$$\mathrm{Hom}_{(\text{as pre-schemes})}(X, Y) \xrightarrow{\sim} \mathrm{Hom}_{(\text{as rings with } 1)}(A, \Gamma(X, \underline{o}_X)) .$$

<u>Corollary</u>: The category of affine schemes (schemes of type Spec (A)) is equivalent to the category of commutative rings with 1, after reversing arrows.

<u>Example</u>: If k is a field, a morphism f: $X \to$ Spec (k) is equivalent to making $\Gamma(X, \underline{o}_X)$ into a k-algebra; or, locally, if X is covered by open sets Spec (A_i), to making each A_i a k-algebra, so that each stalk $\underline{o}_{x,X}$ has a unique k-algebra structure.

<u>Remark</u>: Suppose f: $X \to$ Spec (A) corresponds to the homomorphism φ: $A \to \Gamma(X, \underline{o}_X)$. The map f, as a map of <u>sets</u>, is reconstructed from φ as follows: let $x \in X$, and let φ_x be the composition:

$$A \to \Gamma(X, \underline{o}_X) \to \underline{o}_x .$$

Then f(x) corresponds to the prime ideal

$$\varphi_x^{-1} (m_x) ,$$

where $m_x \subset \underline{o}_x$ is the maximal ideal.

The more classical pre-schemes are characterized as follows:

<u>Proposition-Definition.</u> Let f: $X \to Y$ be a morphism of pre-schemes. Then f is said to be <u>of finite type</u>, if either of the following two equivalent statements is true:

(i) there is an affine open covering U_i = Spec (A_i) of Y, and for each i, there is a <u>finite</u> affine open covering V_{ij} = Spec (B_{ij}) of $f^{-1}(U_i)$ such that for each i, j, B_{ij} as an A_i-algebra of finite type,

(ii) for all affine open sets U = Spec (A) in Y, $f^{-1}(U)$ is quasi-compact (i.e., every open covering admits a finite sub-covering), and for every affine open set $V = \mathrm{Spec}\ (B)$ in $f^{-1}(U)$, B is an A-algebra of finite type.

Definition: Let k be a field, then a pre-scheme X, plus a morphism $f\colon X \to \mathrm{Spec}\ (k)$ is said to be an algebraic pre-scheme /k if f is of finite type. Moreover, if k is algebraically closed, then we will call X a pre-variety /k if X itself is irreducible, and $\underline{o}_X$ has no nilpotent elements ("X is reduced"). This is equivalent to saying that X is covered by affine open sets $\mathrm{Spec}\ (A_i)$, where the A_i are integral domains in the same field K, and the prime ideals $(0) \subset A_i$ all correspond to the same point x of X with stalk $\underline{o}_{x,X} = K$.

3° Since the points of a pre-scheme are so odd, it might be thought that they don't play exactly the same role as points in other geometric theories. (This is true.) It is natural to ask the question: What is the categorical meaning of points? With respect to this question, the category of pre-schemes exhibits significant structural differences from other categories.

Example 1: Let C = category of differentiable manifolds. Let z = the manifold with one point. Then for any manifold X,

$$\mathrm{Hom}_C(z, X) \cong X \text{ as a point set .}$$

Example 2: Let C = category of groups. Let $z = \mathbf{Z}$. Then for any group G,

$$\mathrm{Hom}_C(z, G) \cong G \text{ as a point set .}$$

Example 3: Let C = category of rings with 1 (and homomorphisms f such that $f(1) = 1$). Let $z = \mathbf{Z}[X]$. Then for any ring R,

$$\mathrm{Hom}_C(z, R) \cong R \text{ as a point set .}$$

This indicates that if C is any category, and z is an object, one can try to conceive of $\mathrm{Hom}_C(z, X)$ as the underlying set of points of the object X. In fact:

$$X \longmapsto \mathrm{Hom}_C(z, X)$$

extends to a functor from the category C to the category (Sets), of sets. But, it is not satisfactory to call $\mathrm{Hom}_C(z, X)$ the set of points of X unless this functor is faithful, i.e., unless a morphism f from X_1 to X_2 is determined by the map of sets:

$$\tilde{f}\colon \mathrm{Hom}_C(z, X_1) \to \mathrm{Hom}_C(z, X_2) \quad .$$

Example 4: Let (Hot) be the category of CW-complexes, where $\mathrm{Hom}(X, Y)$ is the set of homotopy-classes of continuous maps from X to Y. If z = the 1 point complex, then

$$\mathrm{Hom}_{(\mathrm{Hot})}(z, X) = \pi_0(X) \text{ (the set of components of } X)$$

and this does not give a faithful functor.

Example 5: Let C = category of pre-schemes. Taking the lead from Examples 1 and 4, take for z the final object of the category C: z = Spec (Z). Now

$$\mathrm{Hom}_C(\mathrm{Spec}\ (\mathbf{Z}),\ X)$$

is absurdly small, and does not give a faithful functor.

Grothendieck's ingenious idea is to remedy this defect by considering not one z, but all z: attach to X the whole set:

$$\bigcup_z \mathrm{Hom}_C(z,\ X)\ .$$

In a natural way, this always gives a faithful functor from the category C to the category (Sets). Even more than that, the "extra structure" on the set $\bigcup_z \mathrm{Hom}_C(z, X)$ which characterizes the object X, can be determined. It consists in:

(i) the decomposition of $\bigcup_z \mathrm{Hom}_C(z, X)$ into subsets $S_z = \mathrm{Hom}_C(z, X)$, one for each z,

(ii) the natural maps from one set S_z to another $S_{z'}$, given for each morphism $g:\ z' \to z$ in the category.

Putting this formally, it comes out like this:

Attach to each X in C, the functor h_X (contravariant, from C itself to (Sets)) via

(*) $h_X(z) = \mathrm{Hom}_C(z, X)$, z an object in C,

(**) $h_X(g)$ = induced map from $\mathrm{Hom}_C(z, X)$ to $\mathrm{Hom}_C(z', X)$ — $g:\ z' \to z$ a morphism in C

Now the functor h_X is an object in a category too: viz,

$$\underline{\mathrm{Funct}}\ (C^{\circ},\ (\mathrm{Sets})),$$

(where Funct stands for functors, C° stands for C with arrows reversed). It is also clear that if $g:\ X_1 \to X_2$ is a morphism in C, then one obtains a morphism of functors $h_g:\ h_{X_1} \to h_{X_2}$. All this amounts to a functor:

$$h:\ C \to \underline{\mathrm{Funct}}\ (C^{\circ},\ \mathrm{Sets})).$$

Proposition: h is fully faithful, i.e., if X_1, X_2 are objects of C, then, under h,

$$\mathrm{Hom}_C(X_1,\ X_2) \xrightarrow{\ \sim\ } \mathrm{Hom}_{\underline{\mathrm{Funct}}}(h_{X_1},\ h_{X_2})\ .$$

Proof: Utterly trivial.

The conclusion, heuristically, is that an object X of C can be identified with the functor h_X, which is basically just a structured set.

The examples of algebraic geometry: If X is a pre-scheme, then morphisms from S to X, i.e., elements of $h_X(S)$, will be referred to as

S-valued points of X

or

S-rational points of X .

For example, very important is the case $S = \operatorname{Spec} \Omega$, Ω an algebraically closed field. Then Ω-valued points of X are called *geometric points* of X (with respect to Ω). The full functor h_X is the absolute functor of points of X. Equally important in algebraic geometry, however, is the relative case—here one fixes a base pre-scheme S (such as Spec (k)), and one looks at the "relativized category":

(*) all objects are pre-schemes X, *plus* structure morphisms $f\colon X \to S$,

(**) all morphisms are morphisms $g\colon X_1 \to X_2$ such that:

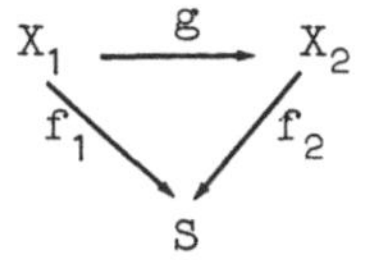

commutes.

(An analogous example is the category of analytic spaces, where $S = \operatorname{Spec}(\mathbf{C})$: a morphism of analytic spaces is required to "pull-back" constant functions to constant functions.)

As a final illustration, we contrast two examples: let C = category of algebraic pre-schemes /k, where k is an algebraically closed field, and let C_0 be the full subcategory of *reduced* algebraic pre-schemes. If $z = \operatorname{Spec}(k)$, then the "points" $h_X(z)$ of an algebraic scheme X are precisely:

i) the closed points of X as a scheme,

ii) the k-valued points of X, as defin d above,

iii) if X is reduced, then the "points" of X in the classical language, e.g., in Serre's FAC.

z is even a final object in the category C. Serre's treatment becomes very simple insofar as $X \mapsto h_X(z)$ is a *faithful* functor so long as one sticks to the subcategory C_0: these pre-schemes may as well be thought of as sets of k-rational points. But $X \mapsto h_X(z)$ is not faithful on C, due to nilpotent elements, and one must look at the whole functor h_X on C.

LECTURE 4

USES OF THE FUNCTOR OF POINTS

1° Grothendieck's Existence Problem: First of all, if $S = \text{Spec}(R)$, an S-valued point of a pre-scheme X will be called simply an R-valued point of X. An R-valued point is simply a generalization of the concept of a solution of a set of equations in R. Thus suppose

$$f_1, \ldots, f_m \in \mathbf{Z}[X_1, \ldots, X_n]$$
$$X = \text{Spec } \mathbf{Z}[X]/(f) .$$

Then one checks immediately that an R-valued point of X is precisely a solution of the equations

$$f_i(\alpha_1, \ldots, \alpha_n) = 0, \qquad 1 \leq i \leq m$$

with $\alpha_j \in R$. The interesting point is that a pre-scheme is actually determined by the functor of its R-valued points as well as by the larger functor of its S-valued points. To state this precisely, if X is a pre-scheme, let $h_X^{(0)}$ be the covariant functor from the category (Rings) of commutative rings with 1 to the category (Sets) defined by:

$$h_X^{(0)}(R) = h_X(\text{Spec } R) = \text{Hom}(\text{Spec } R, X) .$$

Regarding $h_X^{(0)}$ as a functor in X in a natural way, one has:

THEOREM: For any two pre-schemes X_1, X_2,

$$\text{Hom}(X_1, X_2) \xrightarrow{\sim} \text{Hom}(h_{X_1}^{(0)}, h_{X_2}^{(0)}) .$$

Hence $h^{(0)}$ is a fully faithful functor from the category of pre-schemes to

<u>Funct</u> ((Rings), (Sets)).

This result is more readily checked privately than proven formally, but it may be instructive to sketch how a morphism $F: h_{X_1}^{(0)} \to h_{X_2}^{(0)}$ will induce a morphism $f: X_1 \to X_2$. One chooses an affine open covering $U_i \cong \text{Spec}(A_i)$ of X_1; let

$$L_i: \text{Spec}(A_i) \cong U_i \hookrightarrow X_1$$

be the inclusion. Then L_i is an A_i-valued point of X_1. Therefore,

$F(L_i) = f_i$ is an A_i-valued point of X_2, i.e., f_i defines

$$U_i \cong \mathrm{Spec}\,(A_i) \to X_2 .$$

Modulo a verification that these f_i patch together on $U_i \cap U_j$, these f_i give the morphism f via

$$\begin{array}{ccc} U_i & \xrightarrow{\;f_i\;} & X_2 \\ \cap & \nearrow_{f} & \\ X_1 & & \end{array}$$

Grothendieck's existence problem comes up when one asks: Why not <u>identify</u> a pre-scheme X with its corresponding functor $h_X^{(0)}$, and try to define pre-schemes as suitable functors:

$$F\colon (\text{Rings}) \to (\text{Sets}) .$$

The problem is to find "natural" conditions on the functor F to ensure that it is isomorphic to a functor $h_X^{(0)}$. For example, let me mention one property of all the functors $h_X^{(0)}$ which was discovered by Grothendieck: (Compatibility with faithfully flat descent):

Let $q\colon A \to B$ be a homomorphism of rings making B into a faithfully flat A-algebra, i.e.,

(*) $\forall$ ideals $I \subset A$,

$$I \underset{A}{\otimes} B \xrightarrow{\sim} I.B, \quad \text{and} \quad q^{-1}(I.B) = I .$$

Then, if $p_1, p_2\colon B \to B \otimes_A B$ are the homomorphisms $\beta \to \beta \otimes 1$ and $\beta \to 1 \otimes \beta$, the induced diagram of sets:

$$F(A) \xrightarrow{F(q)} F(B) \underset{F(p_2)}{\overset{F(p_1)}{\rightrightarrows}} F(B \underset{A}{\otimes} B)$$

is exact, (i.e., $F(q)$ injective, and $\mathrm{Im}\, F(q) = \{x : F(p_1)x = F(p_2)x\}$).

This approach to the definition of pre-schemes, or of objects in other categories has been used, for example:

(a) by Matsusaka—the theory of Q-varieties is basically an attempt to look at the properties of more general functors F,

(b) by Tate—to give a definition of a <u>global</u> $\wp$-adic analytic space, by a suitable functor F, which can have more structure than a mere local ringed space,

(c) by Murre—in the case of functors from (Rings) to (Groups), where a satisfactory solution to Grothendieck's existence problem seems possible,

(d) by Brown—in the category (Hot), "essentially" all functors turn out to define CW-complexes.

2° Set-theoretic operations lifted to categories: by using the functors h_X, a concept in the thoery of sets may be often defined in arbitrary categories C:

Case 1: "One point." The object X in C which is the analog of "one point" should be that object whose functor satisfies:

$$h_X(S) = \text{a set with one element},$$

for all S. Such an X is, of course, called the "final object" of C.

Case 2: "Group objects" (or, by obvious generalizations, a "ring object," "field object," etc). One can say that X has the structure of a group object in C if

(i) for all S in C, one endows the set $h_X(S)$ with the structure of a group,

(ii) for all $S \xrightarrow{f} S'$, in C, the induced map of sets $h_X(f): h_X(S') \to h_X(S)$ is a homomorphism.

Equivalently, one asks for a lifting of the functor h_X:

$$\begin{array}{ccc} C & \dashrightarrow & (\text{Groups}) \\ & {\scriptstyle h_X}\searrow & \downarrow \\ & & (\text{Sets}) \end{array}$$

If one applies this concept to the category of pre-schemes /S, the objects so defined are called group pre-schemes /S. If $S = \text{Spec}(\mathbf{Z})$, i.e., one considers the category of all pre-schemes, the objects are called absolute group pre-schemes. I will give two examples of such group pre-schemes:

(a) let π be a finite group. Consider the functor F such that:

$$F(S) = \pi,$$

for all connected pre-schemes S (the maps all being the identity from π to π). More generally, one is forced to put:

$$F(S) = \left\{ \begin{array}{l} \text{continuous functions } \alpha \text{ from} \\ S \text{ to } \pi \text{ (with the discrete topology on } \pi) \end{array} \right\}$$

Then F is represented by

$$\pi = \text{Spec}\,(\overbrace{\mathbf{Z} \oplus \mathbf{Z} \oplus \ldots \oplus \mathbf{Z}}^{\text{one copy for each } \sigma}) = \text{Spec}\,(\mathbf{Z}^{\pi})$$

(check this via Theorem 1, Lecture 3), and π is the absolute group scheme corresponding to π.

(b) Work in the category of pre-schemes over $S = \text{Spec}\,(\mathbf{Z}/2)$, i.e., those pre-schemes where $0 = 2$ in $\Gamma(X, \underline{o}_X)$. Consider the functor F defined by:

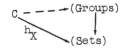

$$F(X) = \{s \in \Gamma(X, \underline{o}_X) \mid s^2 = 1\}$$
$$\text{(a group under multiplication)} \quad .$$

F is, so-to-speak, the set of square roots of 1 in characteristic 2: non-trivial such s certainly exist in rings with nilpotent elements! F is represented by

$$\operatorname{Spec} \{(\mathbf{Z}/2)[X]/(X^2 + 1)\}$$

(check via Theorem 1, Lecture 3).

Case 3: "Hom-objects." Suppose C is a category where products exist (cf. below). Then one can try to lift the *set* Hom (X, Y), for two objects X, Y, into a third *object* $\underline{\text{Hom}}\ (X, Y)$ in C. One method is by the "associativity formula":

$$\text{Hom}\ (S, \underline{\text{Hom}}\ (X, Y)) \cong \text{Hom}\ (S \times X, Y) \ .$$

If one asks for the above isomorphism of the left and right to hold between both sides as *functors in S*, this determines the functor $h_{\underline{\text{Hom}}(X,Y)}$ up to isomorphism, and hence determines the object $\underline{\text{Hom}}\ (X, Y)$.

3° *Fibre products and their uses*: by far the most important categorical notion for algebraic geometry is that of fibre product. One is given a diagram:

$$(*) \qquad \begin{array}{ccc} X & & Y \\ & {\scriptstyle q_1}\searrow \quad \swarrow{\scriptstyle q_2} & \\ & Z & \end{array}$$

If X, Y and Z are sets, then the fibre product is simply

$$X \underset{Z}{\times} Y = \{(x, y) \mid x \in X, y \in Y, q_1(x) = q_2(y)\}.$$

If X, Y and Z are objects in a category C, one can at least form the functor:

$$F(S) = h_X(S) \underset{h_Z(S)}{\times} h_Y(S) \ .$$

If $h_W \cong F$, then W is written $X \times_Z Y$, and called the fibre product. One checks that to find W is the same thing as completing (*) to:

$$\begin{array}{ccc} & X \underset{Z}{\times} Y & \\ {\scriptstyle p_1}\swarrow & & \searrow{\scriptstyle p_2} \\ X & & Y \\ {\scriptstyle q_1}\searrow & & \swarrow{\scriptstyle q_2} \\ & Z & \end{array} \qquad \text{(commuting)}$$

with the universal mapping property:

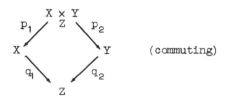

(UMP) for all objects S, and morphisms $S \xrightarrow{f} X$, $S \xrightarrow{g} Y$, such that $q_1 \circ f = q_2 \circ g$, there is a unique morphism $S \xrightarrow{h} X \times_Z Y$ such that $f = p_1 \circ h$, $g = p_2 \circ h$. This h will be written $(f, g)_Z$ or (f, g). The notations p_1 and p_2 will always be used for the projections of $X \times_Z Y$. The basic result is:

THEOREM: In the category of pre-schemes, fibre products always exist. (Cf. EGA, Ch. 1, §3.) This should be used in conjunction with the more precise result:

$$\operatorname{Spec} A \underset{\operatorname{Spec} C}{\times} \operatorname{Spec} B \cong \operatorname{Spec} (A \underset{C}{\otimes} B)\ ,$$

and the fact that, if $U \subset X$ and $V \subset Y$ are open subsets, then $U \times_S V$ is an open subset of $X \times_S Y$. The proof of both results is very easy. Knowing what fibre products are, we can define many operations and concepts:

Application 1: Field extension—as in classical algebraic geometry. Let $k \subset K$ be two fields, and let X be an algebraic pre-scheme over k. To consider the "same" X over the larger field K, one forms the fibre product:

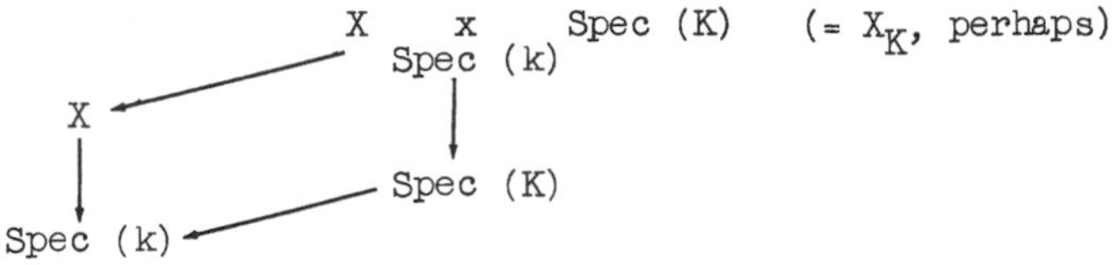

For example, suppose $K = \Omega$ is algebraically closed. As an application, we prove:

Proposition. The following sets are canonically isomorphic:

(i) the geometric points of X (with respect to Ω),

(ii) $\operatorname{Hom}_{\operatorname{Spec}(k)} (\operatorname{Spec} \Omega, X)$,

(iii) points $x \in X$, plus k-injections

$$\mathcal{K}(x) \hookrightarrow \Omega \quad (\mathcal{K}(x) = \text{residue field of } \underline{o}_x),$$

(iv) $\operatorname{Hom}_{\operatorname{Spec} \Omega} (\operatorname{Spec} \Omega, X_\Omega)$,

(v) closed points of X_Ω.

Proof: (i) and (ii) are equal by definition. Their equality with (iii) follows immediately from the definition of a morphism in a local ringed space. The equality of (ii) and (iv) results from the definition of the fibre product X_Ω. To check the equality of (iv) and (v), we may assume X_Ω is the affine scheme Spec (A), where A is a finitely generated algebra over Ω, then

$$\operatorname{Hom}_{\operatorname{Spec} \Omega} (\operatorname{Spec} \Omega, X_\Omega) = \operatorname{Hom}_\Omega(A, \Omega)$$

and one uses the well-known:

(*) if $m \subset A$ is a maximal ideal, then $A/m \cong \Omega$.

QED

<u>Application 2</u>: Fibres of a morphism. Let $f: X \to Y$ be a morphism of pre-schemes, and $y \in Y$ be any point. Let $\kappa(y)$ = residue field of $\underline{o}_y$: y determines a canonical morphism:

$$\operatorname{Spec} \kappa(y) \xrightarrow{i_y} Y$$

via $\left\{\begin{array}{l} \text{the pt.} \to y \\ \kappa(y) \leftarrow \underline{o}_y \quad \text{(canonically)} \end{array}\right.$

One forms the fibre product:

$$\begin{array}{ccc} X & \longleftarrow & X \times_Y \operatorname{Spec} \kappa(y) = \underline{f^{-1}(y)}, \text{ or } \underline{X_y}\,. \\ {\scriptstyle f}\big\downarrow & & \big\downarrow \\ Y & \xleftarrow{\ i_y\ } & \operatorname{Spec} \kappa(y) \end{array}$$

This is the scheme-theoretic fibre of f. Similarly, if $g: \operatorname{Spec} \Omega \to Y$ is a <u>geometric</u> point of Y, then the fibre product:

$$\begin{array}{ccc} X & \longleftarrow & X \times_Y \operatorname{Spec} \Omega \\ {\scriptstyle f}\big\downarrow & & \big\downarrow \\ Y & \xleftarrow{\ g\ } & \operatorname{Spec} \Omega \end{array}$$

is called the geometric fibre of f over the given geometric point. In this language, one has the droll result:

<u>Proposition</u>: Let $k \subset K$ be two fields, and let $f: \operatorname{Spec} K \to \operatorname{Spec} k$ be induced by the inclusion of k in K. Then,

$$[K/k \text{ is separable}] \Longleftrightarrow \left[\begin{array}{l} \text{one (and hence all) geometric fibres} \\ \text{of } f \text{ are reduced schemes} \end{array}\right]$$

(Proof is left to reader.)

<u>Application 3</u>: Direct definition of a group pre-scheme /S. After all, a group is simply a set X plus three maps:

$$\begin{array}{ll} \text{mult:} & X \times X \to X \\ \text{inverse:} & X \to X \\ \text{identity:} & \{e\} \to X \end{array}$$

satisfying well-known relations. Therefore, a group pre-scheme X/S consists in the functor h_X (on the category of pre-schemes /S) plus three morphisms of functors:

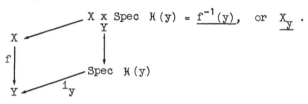

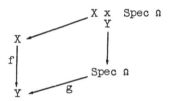

$$\begin{array}{ll} \text{mult:} & h_X \times h_X \to h_X \\ \text{inverse:} & h_X \to h_X \\ \text{identity:} & \\ \{1 \text{ elt. functor}\} \to h_X & \end{array}$$

satisfying the same identities. But: (a) $h_X \times h_X$ is isomorphic to $h_{X \times_S X}$, and (b) {1 elt. functor} is isomorphic to h_S, S being the final object in our category. Therefore, X is a group pre-scheme /S if one is given three morphisms:

$$\begin{array}{ll} \text{mult:} & X \times_S X \to X \\ \text{inverse:} & X \to X \\ \text{identity:} & S \to X \end{array}$$

satisfying the same identities.

A final point not to be forgotten: if X is a group pre-scheme /S, for all T/S, the T-valued points of X form a group: but in no sense do the ordinary points of X form a group (even if $S = \operatorname{Spec} \Omega$).

Application 4: Definition of a scheme. Let X be a pre-scheme, and let 1_X: $X \to X$ be the identity. The induced morphism

$$\Delta = (1_X, 1_X): X \to X \times X$$

is called the diagonal.

Proposition-Definition: X is a scheme if equivalently:

i) $\Delta(X)$ is closed in $X \times X$,

ii) for every pair of morphisms $Y \overset{f_1}{\underset{f_2}{\rightrightarrows}} X$,

$$\{y \in Y \mid f_1(y) = f_2(y)\} \text{ is a closed subset of } Y$$

Proof: ii) ===> i) by taking $Y = X \times X$, $f_i = i^{\text{th}}$ projection p_i of $X \times X$ on X; i) ===> ii) by factoring f_i:

$$Y \xrightarrow{(f_1, f_2)} X \times X \overset{p_1}{\underset{p_2}{\rightrightarrows}} X,$$

and noting that

$$\{y \in Y \mid f_1 y = f_2 y\} = (f_1, f_2)^{-1}[\Delta(X)] .$$

QED

From now on, we will deal only with schemes, unless otherwise specified.

APPENDIX TO LECTURE 4

RE REPRESENTABLE FUNCTORS AND ZARISKI TANGENT SPACES

As an application both of the concepts of functors and of nilpotents, we connect these to the geometric concept of the Zariski tangent space. Assume that X is a scheme over a field k, and that $x \in X$ is a k-rational point, i.e., the given homomorphism $k \to \underline{o}_x$ induces an isomorphism $k \xrightarrow{\sim} \mathcal{K}(x)$.

Definition: If $m \subset \underline{o}_x$ is the maximal ideal, then the dual vector space to m/m^2 is the Zariski tangent space T_x to X at x.

Now consider the interesting class of schemes:

Definition: If V is a vector space (always finite dimensional) over k, let

$$I_V = \text{Spec } (k \oplus V) ,$$

where $k \oplus V$ is a ring via $V^2 = (o)$. Note that one has two homomorphisms:

$$k \rightleftarrows k \oplus V$$

(via $\alpha \mapsto \alpha + 0$; $\alpha + v \mapsto \alpha$), hence two morphisms:

$$\text{Spec } k \underset{j}{\overset{\longleftarrow}{\longrightarrow}} I_V \ .$$

We work entirely in the category of schemes and morphisms over Spec (k). Now suppose $f\colon I_V \to X$ is any morphism over Spec (k). I_V has only one point, and its image under f must be a k-rational point $x \in X$. Then f is determined by x, and by a local k-homomorphism:

$$\underline{o}_x \xrightarrow{f^*} k \oplus V \ .$$

But f^* is just a linear map from m/m^2 to V, i.e., an element of $V \otimes_k T_x$. This gives:

Proposition: For all schemes $X/\text{Spec }(k)$, there is a natural isomorphism between

$\text{Hom}_k(I_V, X)$ and $\{k\text{-rational pts } x \in X, \text{ plus elements of } V \otimes_k T_x\}$.

In particular, regarding k as a 1-dimensional vector space over itself, the subset of $\text{Hom}_k(I_k, X)$ with given image x is isomorphic to the tangent space T_x itself, i.e., the Zariski-tangent-space can be recovered from the set of I_k-valued points of X.

In fact, even the vector-space structure on the set of I_k-valued points with given image can be defined directly in terms of the functor of points of X. More than that, there is a very general class of contravariant functors F (from schemes /k to (sets)) for which one can in the same way define Zariski tangent spaces, even though they may not be representable.

To see this, fix such a functor F. Then the set F(Spec (k)) is the set of k-rational points x of F. Fix one such x. For all vector spaces V, the subset of $F(I_V)$ "whose image point is x" can be interpreted as:

$$F(I_V)_x = \{\xi \in F(I_V) \mid j^*(\xi) = x \text{ in } F(\mathrm{Spec}\,(k))\}\ .$$

(where j: Spec (k) $\to I_V$ is the morphism defined above). I claim that for "reasonable" functors F, the set $F(I_k)_x$ has a canonical structure of vector space and that this is the tangent space to F at x! The property F must have is:

(*) for all vector spaces V_1, V_2:

$$F(I_{V_1 \oplus V_2})_x \xrightarrow{\sim} F(I_{V_1})_x \times F(I_{V_2})_x$$

[where the map is given by the projections $V_1 \oplus V_2 \to V_i$ which induce morphisms $I_{V_i} \to I_{V_1 \oplus V_2}$, hence maps $F(I_{V_1 \oplus V_2})_x \to F(I_{V_i})_x$].

Assuming this, fix $\xi_1, \xi_2 \in F(I_k)_x$, and $\alpha, \beta \in k$. What is $\alpha\xi_1 + \beta\xi_2$? Well, use the diagram:

$$F(I_k)_x \times F(I_k)_x \xleftarrow{\sim} F(I_{k \oplus k})_x \xrightarrow{[\alpha,\beta]} F(I_k)_x$$

where $[\alpha, \beta]$ is induced by the homomorphism $(\gamma, \delta) \to (\alpha\gamma + \beta\delta)$ from $k \oplus k$ to k. The image of $\xi_1 \times \xi_2$ is defined to be $\alpha \cdot \xi_1 + \beta \cdot \xi_2$. We leave it as an exercise to check that this does make $F(I_k)_x$ into a vector space.

LECTURE 5

Proj AND INVERTIBLE SHEAVES

So far, the only schemes which we have constructed have been affine schemes Spec (R). We now introduce a second fundamental construction Prof (R) which attaches to a graded ring:

$$R = \sum_{n=0}^{\infty} R_n$$

a scheme which is almost never affine.

$X = \text{Proj}\,(R) = $

a) as point set, the set of homogeneous prime ideals $p \subset R$, such that

$$p \not\supset \sum_{n=1}^{\infty} R_n ,$$

b) as topological space, a basis of open sets is given by the subsets

$$X_f = \{p \mid f \notin p\}, \text{ for } f \in R_n, \ n > 0,$$

c) as local ringed space, its structure sheaf is defined via:

$$\Gamma(X_f, \underline{o}_X) = [R_{(f)}]_{(0)}$$
$$= \text{subring of } R_{(f)} \text{ of elts. of degree } 0.$$

Proposition 1. X is a scheme (n.b. not just a pre-scheme).

High Points of Proof: One shows that

$$X_f \cong \text{Spec}\,[R_{(f)}]_{(0)}$$

by mapping a homogeneous prime $p \subset R$ (such that $f \notin p$) to $p \cdot R_{(f)} \cap [R_{(f)}]_{(0)}$; the topologies correspond in virtue of

$$X_f \cap X_g = X_{f\cdot g} = [\text{open subset of } X_f \text{ defined by } (\frac{g^m}{f^n}) \neq 0]$$

where $f \in R_m$, $g \in R_n$.

The most important Proj is:

$$\mathbf{P}_n = \text{Proj}\ \mathbf{Z}[X_0, X_1, \ldots, X_n].$$

Incidentally, the actual "appearance" of $\mathbf{P}_1$ can be described somewhat-we have divided up points via the dimension of their stalks, and via their images in Spec ($\mathbf{Z}$); also the closure of $1/5$ and $\sqrt{-1}$ are "illustrated":

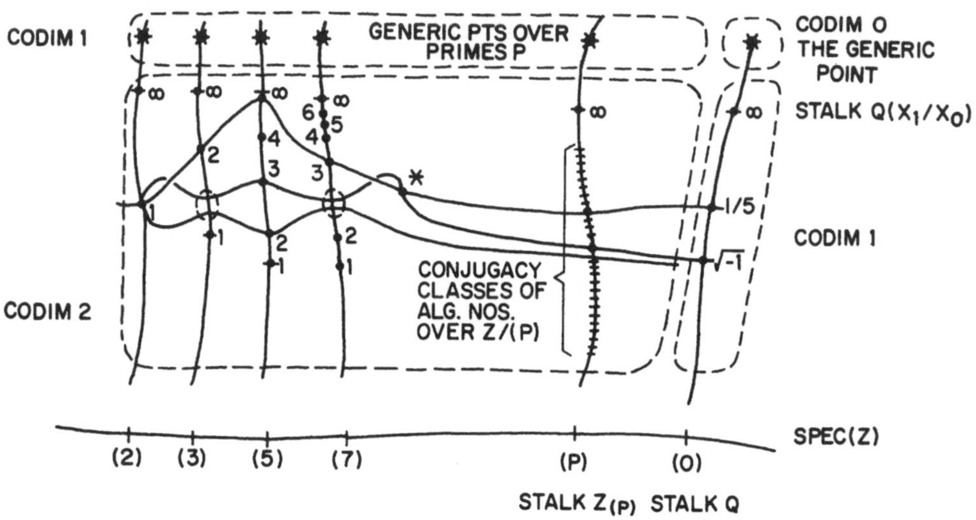

Exercise: What is the point (*)?

A more weighty question is what are the S-valued points in $\mathbf{P}_n$, i.e., what is the functor of $h_{\mathbf{P}_n}$. The answer to this question involves us immediately in a new concept.:

Definition: If X is a local ringed space, a sheaf $\mathfrak{L}$ of $\underline{o}_X$-modules such that there exists a covering $\{U_i\}$ of X for which

$$\mathfrak{L}|_{U_i} \cong \underline{o}_X|_{U_i} \quad ,$$

as $\underline{o}_X$-modules,

is called an invertible sheaf.

More concretely, what is such an $\mathfrak{L}$? Since locally it is isomorphic to $\underline{o}_X$, the essential part of $\mathfrak{L}$ is in the patching: i.e., $\mathfrak{L}$ can be constructed by starting with $\underline{o}_X$ on each U_i, and patching these as sheaves of $\underline{o}_X$-modules on $U_i \cap U_j$. But

$$\mathrm{Hom}_{\substack{\text{as sheaves of}\\ \underline{o}_X\text{-modules}}}(\underline{o}_X|_{U_i \cap U_j}, \underline{o}_X|_{U_i \cap U_j}) \cong \Gamma(U_i \cap U_j, \underline{o}_X)$$

[where $h \in$ Hom corresponds to $h(1) \in \Gamma(U_i \cap U_j, \underline{o}_X)$; and $f \in \Gamma(U_i \cap U_j, \underline{o}_X)$ corresponds to the homomorphism given by multiplication by f]. Now define:

Definition: An element $s \in \Gamma(U, \underline{o}_X)$ is a unit if equivalently:

i) there exists a multiplicative inverse $s^{-1} \in \Gamma(U, \underline{o}_X)$

or ii) for all $x \in U$, the induced element s_x in $\underline{o}_x$ is not in the maximal ideal m_x.

It is clear from (ii) that the units form a subsheaf of $\underline{o}_X$ -which we will denote $\underline{o}_X^*$. It is clear from (i) that $\underline{o}_X^*$ is a sheaf of groups under multiplication. Now it is clear that the isomorphisms of $\underline{o}_X$ with itself are:

$$\mathrm{Isom}_{\substack{\text{as sheaves of}\\ \underline{o}_X\text{-modules}}}(\underline{o}_X|_{U_i\cap U_j}, \underline{o}_X|_{U_i\cap U_j}) \cong \text{units in } \Gamma(U_i \cap U_j, \underline{o}_X)$$

Therefore, to construct $\mathfrak{L}$, $\underline{o}_X$ must be patched to itself on $U_i \cap U_j$ by multiplication by a unit s_{ij} over $U_i \cap U_j$. Since all these identifications must be compatible on $U_i \cap U_j \cap U_k$, it follows that:

$$s_{ij} \cdot s_{jk} \cdot s_{ki} = 1 \quad \text{on} \quad U_i \cap U_j \cap U_k .$$

This means that $\{s_{ij}\}$ form a 1-Czech co-cycle, and we have defined an element λ of $H^1(X, \underline{o}_X^*)$. The main, but elementary, result in this direction is:

<u>Proposition 2</u>: λ depends only on $\mathfrak{L}$, and this sets up an isomorphism between the set of invertible sheaves on X-modulo isomorphism—and the set $H^1(X, \underline{o}_X^*)$.

<u>Definition</u>: $\mathrm{Pic}(X) = H^1(X, \underline{o}_X^*)$.

<u>Remarks</u>: A) Pic (X) is a commutative group—this is clear since $\underline{o}_X^*$ is a sheaf of groups. More directly, if $\mathfrak{L}_1$ and $\mathfrak{L}_2$ are invertible sheaves, their product is $\mathfrak{L}_1 \otimes \mathfrak{L}_2$; and if $\mathfrak{L}_1$ is given by the co-cycle s_{ij} with respect to $\{U_i\}$, and L_2 is given by t_{ij} for the <u>same</u> covering, then $\mathfrak{L}_1 \otimes \mathfrak{L}_2$ is simply the sheaf given by the patching $s_{ij} \cdot t_{ij}$.

B) Pic (X) is a contravariant functor with respect to X. Given any $X \xrightarrow{f} Y$, there is a homomorphism $\underline{o}_Y^* \xrightarrow{f^*} \underline{o}_X^*$, hence an induced homomorphism of H^1's. More directly, if $\mathfrak{L}$ is an invertible sheaf on Y, then $f^*(\mathfrak{L}) = \underline{o}_X \otimes_{\underline{o}_Y} \mathfrak{L}$ is an invertible sheaf on X; and if $\mathfrak{L}$ is given by the co-cycle s_{ij} with respect to $\{U_i\}$, then $f^*(\mathfrak{L})$ is given by the co-cycle $f^*(s_{ij})$ with respect to $\{f^{-1}(U_i)\}$. Note also that sections

$$s \in \Gamma(Y, \mathfrak{L})$$

induce sections

$$f^*(s) \in \Gamma(X, f^*(\mathfrak{L})) .$$

C) Suppose s is a section of an invertible sheaf $\mathfrak{L}$ on X. Then although s does not have values at points $x \in X$, it does make sense to say <u>$s(x) = 0$</u> or <u>$s(x) \neq 0$</u>. Namely, if we choose an isomorphism of $\mathfrak{L}_x$ and $\underline{o}_x$, and if s corresponds to $g \in \underline{o}_x$, then at least whether the value $g(x) \in \mathcal{H}(x)$ of g is 0 or not is independent of the isomorphism. In particular, one has the subset of X:

$$X_s = \{x \in X \mid s(x) \neq 0\}$$

which is easily seen to be open. These open sets include as special cases the open sets X_f used to define the topology both of Spec (A) and Proj (R) (cf. below (iv)).

Returning to Proj (R), assume that:

(*) R_n is spanned, as R_0-module, by $\overbrace{R_1 \otimes \ldots \otimes R_1}^{nx}$.

Then we find that Proj (R) has more structure:

i) $X = \text{Proj}(R)$ is covered by X_f, for $f \in R_1$.

[Proof: If $x \in X - \bigcup X_f$, then x corresponds to a $p \subset R$ such that all f in R_1 are in p; Thus $R_1 \subset p$, and $\Sigma_1^\infty R_n \subset p$, contrad.]

ii) On $X_f \cap X_g$, f/g is a unit. Therefore the covering $\{X_f\}$ and the units f/g define a 1-Czech co-cycle on Proj (R), hence an invertible sheaf. This is called $\underline{o}(1)$.

iii) If $\underline{o}(n)$ is the nth tensor power $\underline{o}(1)^{\otimes n}$ of $\underline{o}(1)$, one has a canonical homomorphism

$$R_n \xrightarrow{\varphi_n} \Gamma(X, \underline{o}(n))$$

which is the geometric significance of the graded ring R.

[Construction: $\underline{o}(n)$ is defined by the co-cycle $(f/g)^n$ for the covering $\{X_f\}$. If $k \in R_n$, then k gives rise to the sections k/f^n of $\underline{o}_X$ on X_f; since these differ precisely by factors $(f/g)^n$ on $X_f \cap X_g$, they patch up as sections of $\underline{o}(n)$.]

iv) One checks that, for $k \in R_n$, the open sets X_k defining the topology on $X = \text{Proj}(R)$ are the same as the open sets $X_{\varphi_n(k)}$ defined as in C) above.

Let us apply this new information to study the structure of the functors $h_{\text{Proj}(R)}$. Given an S-valued point

$$S \xrightarrow{f} \text{Prof}(R)$$

of Proj (R), one obtains on S an induced invertible sheaf $f^*(\underline{o}(1))$ on S. Putting this functorially, one has a very basic morphism of functors:

$$h_{\text{Proj}(R)} \rightarrow \text{Pic} .$$

This is interesting from two standpoints: it explains the non-triviality of the functor of points of a Proj; and it is a beginning in representing the functor Pic. Although it may seem strange to view Proj (R), or $\mathbf{P}_n$, as approximate group-schemes, really representing Pic, this is quite accurate in the category (Hot). Here we have the CW-complex $\mathbf{C}\,\mathbf{P}_n$ (complex projective n-space) and

$$\mathbf{C}\,\mathbf{P}_n \hookrightarrow \mathbf{C}\,\mathbf{P}_\infty ,$$

hence

$$\begin{bmatrix}\text{functor represented}\\ \text{by } \mathbf{C}\,\mathbf{P}_n\end{bmatrix} \rightarrow \begin{bmatrix}\text{functor represented}\\ \text{by } \mathbf{C}\,\mathbf{P}_\infty\end{bmatrix}$$

$$\wr\|$$

$$\begin{bmatrix}\text{functor}\\ S \rightarrow H^2(S, \mathbf{Z})\end{bmatrix}$$

$$\wr\|$$

$$\begin{bmatrix}\text{functor}\\ S \rightarrow \text{group of topological equiv. classes of line bundles on } S\end{bmatrix}$$

via $\mathbf{C}\,\mathbf{P}_\infty \cong$ Eilenberg-Maclane Space $K(\mathbf{Z}, 2)$.

We can now give the explicit description of the functor $h_{\mathbf{P}_n}$ which we have been driving at. Let

$$X_i \in \Gamma(\mathbf{P}_n, \underline{o}(1))$$

correspond as in (iii) to X_i in the R_1-component of $\mathbf{Z}[X_0, \ldots, X_n]$. Then for all $S \rightarrow \mathbf{P}_n$, one obtains:

$$\begin{aligned} \mathfrak{L} &= f^*(\underline{o}(1)) \\ s_i &= f^*(X_i) \in \Gamma(S, \mathfrak{L}) \end{aligned}$$

Proposition 3: This gives an isomorphism:

$$h_{\mathbf{P}_n}(S) \xrightarrow{\sim} \left\{ (\mathfrak{L}; s_0, \ldots, s_n) \;\middle|\; \begin{array}{l} \mathfrak{L} \text{ an invertible sheaf on } S \\ s_0, \ldots, s_n \text{ sections of } \mathfrak{L} \\ \text{such that for all } x \in S, \\ \text{there is an } i \text{ such that} \\ s_i(x) \neq 0 \end{array} \right\} \Big/ \begin{array}{l}\text{modulo}\\ \text{isomor-}\\ \text{phism.}\end{array}$$

Proof: Not a difficult exercise, (cf. EGA 2, §4); $f: S \rightarrow \mathbf{P}_n$ is given by a collection $f_i: S_{s_i} \longrightarrow (\mathbf{P}_n)_{X_i}$, $0 \leq i \leq n$, which patch together; since $(\mathbf{P}_n)_{X_i}$ is affine, use Theorem 1, Lecture 3.

A nice Corollary ties this in with the elementary definition of Proj. space over a field k—except we may as well at least replace k by a local ring $\underline{o}$:

Corollary: If $\underline{o}$ is a local ring, the set of $\underline{o}$-valued points of $\mathbf{P}_n$ is isomorphic to:

$$\frac{\{(\alpha_0, \ldots, \alpha_n) \mid \alpha_i \in \underline{o}, \text{ not all } \alpha_i \text{ in the max. ideal } m\}}{(\alpha_0, \ldots, \alpha_n) \sim (\lambda\alpha_0, \ldots, \lambda\alpha_n), \text{ all units } \lambda \in \underline{o}^*}.$$

Proof: Since Spec ($\underline{o}$) itself is the only open subset of Spec ($\underline{o}$) containing the one closed point, it follows that Spec ($\underline{o}$) has only one invertible sheaf, $\underline{o}_{\text{Spec}(\underline{o})}$. Since the automorphisms of $\underline{o}_{\text{Spec}(\underline{o})}$ are precisely multiplications by units $\lambda \in \underline{o}^*$, the Corollary is a special case of Proposition 3.

As a final point, it is interesting to give the generalization of this last Proposition to Grassmannians. Before defining the actual Grassmannian explicitly, we can characterize it by giving its functor:

<u>Definition</u>: A sheaf $\mathcal{E}$ of $\underline{o}_X$-modules is <u>locally free of rank</u> r if there exists an open covering $\{U_i\}$ of X such that

$$\mathcal{E}|_{U_i} \cong \underline{o}_X^r|_{U_i} .$$

Then the functor is:

$$S \mapsto \left\{ \begin{array}{l} \text{locally free sheaves } \mathcal{E} \text{ of rank } r \text{ on } S\text{; plus } (n+1)\text{-sections} \\ s_0, s_1, \ldots, s_n \text{ of } \mathcal{E} \text{ which } \underline{\text{generate}} \ \mathcal{E}, \text{ i.e.,} \\ \qquad \mathcal{E}_x = \sum_{i=0}^{n} \underline{o}_x \cdot s_i, \text{ all } x \in S \end{array} \right\} \Big/ \text{modulo isomorphism}$$

and the embedding in projective space via Plücker coordinates corresponds to the functorial map:

$$\{ \mathcal{E}; s_0, \ldots, s_n\} \mapsto \{\Lambda^r\mathcal{E} ; \ldots, s_{i_1}\wedge\ldots\wedge s_{i_r}, \ldots\}.$$

↑ one for each $0 \leq i_1 < i_2 < \ldots < i_r \leq n$

S-valued pt. of Grassmannian $\mapsto$ S-valued pt. of a projective space.

Let $p_{i_1,\ldots,i_r} = s_{i_1}\wedge\ldots\wedge s_{i_r}$ and $\mathcal{L} = \Lambda^r\mathcal{E}$. Then the sections $p_{i_1,\ldots,i_r}$ satisfy the well-known quadratic relations

$$(\#) \qquad \sum_{\lambda=1}^{r+1} (-1)^{\lambda}\, p_{i_1,i_2,\ldots,i_{r-1},j_\lambda} \otimes p_{j_1,j_2,\ldots,\hat{j}_\lambda,\ldots,j_{r+1}} = 0$$

for any sequences $i_1,\ldots, i_{r-1}$ and $j_1,\ldots, j_{r+1}$.

THEOREM: The above morphism from the Grassmannian functor to the functor of projective space is injective and its image consists precisely of the S-valued points of projective space satisfying (#).

<u>Proof</u>: An S-valued point of the Grassmannian can be regarded as a surjective homomorphism:

$$\underline{o}_S^{n+1} \xrightarrow{\ \varphi\ } \mathcal{E} \to 0 .$$

Up to isomorphism, this point is determined by the kernel of φ; since the kernel is a subsheaf of a fixed sheaf, if it is given locally, it is determined globally. Therefore the result follows if, given any S-valued point of projective space satisfying (#), there is an open covering of S such that over each open subset, the S-valued point lifts uniquely to a point of the Grassmannian. Therefore, we can pass to an open set where a fixed Plücker coordinate

$$p_{i_1,i_2,\ldots,i_r} \neq 0,$$

i.e., this p generates $\mathcal{L}$ globally. The relations (#) can then be "solved," and one checks that they take precisely the form

$$p_{j_1,\ldots,j_r} = \frac{F(\ldots,p_{i_1,\ldots,\hat{i}_k,\ldots,i_r,j},\ldots)}{(p_{i_1,\ldots,i_r})^{N-1}}$$

where at least two of the j's are not in the set $i_1,\ldots, i_r$ and where F is a homogeneous polynomial of degree N in the $r(n+1 - r)$ free variables $p_{i_1,\ldots,\hat{i}_k,\ldots,i_r,j}$. On the other hand, for the S-valued point φ of the Grassmannian functor to induce a projective point where $p_{i_1,\ldots,i_r} \neq 0$, it is necessary and sufficient that $s_{i_1} = \varphi(e_{i_1}),\ldots, s_{i_r} = \varphi(e_{i_r})$ is a basis of the sheaf $\mathcal{E}$. Then the ideal which is the kernel of φ has a unique basis of the form:

$$\left[e_j - \sum_{k=1}^{r} a_{jk}\, e_{i_k}\right]_{j \in \{0,1,\ldots,n\}-\{i_1,\ldots,i_r\}}$$

(where $e_0,\ldots, e_n$ is the standard basis of $\underline{o}_X^{n+1}$). In terms of a_{jk}, the Plücker coordinates come out:

$$a_{jk} = (-1)^{r-k}\,\frac{p_{i_1,\ldots,\hat{i}_k,\ldots,i_r,j}}{p_{i_1,i_2,\ldots,i_r}} .$$

Therefore there is one and only one choice of $a_{jk} \in \Gamma(S, \underline{o}_S)$ corresponding to the given Plücker coordinates.

QED

Corollary 1: The Grassmannian functor is represented by

$$G_{n,r} = \mathrm{Proj}\ \mathbf{Z}[\ldots, p_{i_1,\ldots,i_r},\ldots]/(\text{Quadratic relations}_{\#}).$$

Corollary 2: The open set of $G_{n,r}$ where $p_{i_1,\ldots,i_r} \neq 0$ is isomorphic to affine space of dimension $r(n+1 - r)$.

APPENDIX TO LECTURE 5

A further development of the theory reveals that the operation Proj, as defined above, is often too special. To understand the generalization let $R = \sum_{n=0}^{\infty} R_n$ be a graded ring. Suppose R_0 happens to be an S-algebra; as such it gives a quasi-coherent sheaf

$$\mathbf{R} = \sum_{n=0}^{\infty} \mathbf{R}_n$$

$$\mathbf{R} = \tilde{R}, \quad \mathbf{R}_n = \tilde{R}_n$$

of $\underline{o}_X$-modules on $X = \mathrm{Spec}\ (S)$. Here $\mathbf{R}$ is actually a *quasi-coherent graded sheaf of $\underline{o}_X$-algebras* (a mouthful, but simple enough). The point is that one can encounter such sheaves even on non-affine schemes X. Thus say $\mathbf{R} = \sum_{n=0}^{\infty} \mathbf{R}_n$ is such a creature on some scheme X. Then for all affine open $U \subset X$,

$$\Gamma(U, \mathbf{R}) = \sum_{n=0}^{\infty} \Gamma(U, \mathbf{R}_n)$$

is a graded ring over $\Gamma(U, \underline{o}_X)$. Therefore one gets a scheme $\mathrm{Proj}[\Gamma(U, \mathbf{R})]$, together with a morphism

$$\pi : \mathrm{Proj}\ \Gamma(U, \mathbf{R}) \longrightarrow U .$$

One checks (cf. EGA, 2, §3) that these patch together canonically to a scheme $\underline{\mathrm{Proj}}\ (\mathbf{R})$ together with a morphism:

$$\pi : \underline{\mathrm{Proj}}\ (\mathbf{R}) \longrightarrow X .$$

The following is the most important example: Let E be a locally free sheaf of rank r on a scheme X. Put $\mathbf{R}_n$ equal to the nth symmetric power of E (as $\underline{o}_X$-modules), and $\mathbf{R} = \sum \mathbf{R}_n$. Then one writes:

$$\mathbf{P}(E) = \underline{\mathrm{Proj}}\ (\mathbf{R}) .$$

This scheme generalizes $\mathbf{P}_n$ itself: i.e.,

$$\mathbf{P}_n = \mathbf{P}\left[\bigoplus_{i=0}^{n} X_i \cdot \underline{o}_{\mathrm{Spec}\ \mathbf{Z}} \right] .$$

On the other hand, it is not much more complicated than $\mathbf{P}_n$, for if E is isomorphic to the free sheaf $(\underline{o}_X)^r$ on the open covering $\{U_i\}$ of X, then <u>over</u> U_i:

$$\begin{aligned} \mathbf{P}(E)|_{U_i} &\cong \mathbf{P}((\underline{o}_X)^r)|_{U_i} \\ &\cong \mathbf{P}_{r-1} \times U_i \quad . \end{aligned}$$

[This follows from the general fact that if $f\colon X \to Y$ is any morphism, and $\mathbf{R}$ is a quasi-coherent graded sheaf of $\underline{o}_Y$-algebras, then:

$$\underset{\sim}{\text{Proj}}\,(f^*(\mathbf{R})) \cong \underset{\sim}{\text{Proj}}(\mathbf{R}) \underset{Y}{\times} X \ .$$

cf. EGA 2. §3.5.]

For $\mathbf{P}(E)$, $\underline{o}(1)$ is constructed exactly as before, and one finds a canonical homomorphism:

$$E \to \pi_*(\underline{o}(1))$$

(if π is the projection from $\mathbf{P}(E)$ onto the base X). Moreover, the induced homomorphism on $\mathbf{P}(E)$:

$$\pi^*(E) \to \underline{o}(1)$$

is surjective. Now suppose a morphism $g\colon S \to X$ is given. Then to any lifting h:

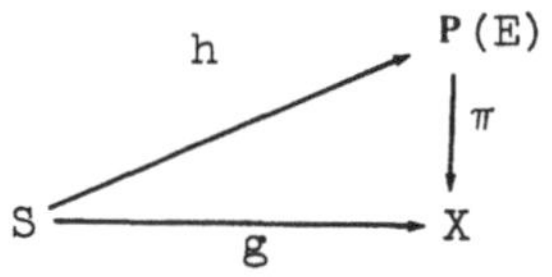

we can associate the invertible sheaf $L = h^*(\underline{o}(1))$, and a surjective homomorphism:

$$\varphi\colon \ g^*(E) = h^*(\pi^* E) \to h^*(\underline{o}(1)) = L \ .$$

An easy generalization of the result for $\mathbf{P}_n$ states that this sets up a functorial isomorphism between the set of S-valued points h of $\mathbf{P}(E)$ lifting g, and the set of L and φ.

LECTURE 6

PROPERTIES OF MORPHISMS AND SHEAVES

1° Affine concepts: Let $X = \mathrm{Spec}\,(R)$. We recall that for all R-modules, M, one can define a sheaf $\widetilde{M}$ of $\underline{o}_X$-modules, via:

$$\Gamma(X_f, \widetilde{M}) = M_{(f)}, \quad \text{all} \quad f \in R .$$

This defines a fully faithful and exact functor:

$$\begin{bmatrix}\text{Category of} \\ \text{R-modules}\end{bmatrix} \rightarrow \begin{bmatrix}\text{Category of sheaves} \\ \text{of } \underline{o}_X\text{-modules}\end{bmatrix}$$

[i.e., $\mathrm{Hom}_{\underline{o}_X}(\widetilde{M}, \widetilde{N}) \cong \mathrm{Hom}_R(M, N)$, and $0 \to \widetilde{M} \to \widetilde{N} \to \widetilde{P} \to 0$ is exact if $0 \to M \to N \to P \to 0$ is exact].

Definition: A sheaf $\mathcal{F}$ of $\underline{o}_X$-modules is quasi-coherent if $\mathcal{F}$ is isomorphic to $\widetilde{M}$, for some R-module M.

Example: Let R be a discrete, rank 1 valuation ring with quotient field K. Then there are two nonempty open sets in Spec(R): the whole space X, and the generic point itself U. A sheaf $\mathcal{F}$ of $\underline{o}_X$-modules consists, therefore, in

a) an R-module $A = \mathcal{F}(X)$; a K-vector space $B = \mathcal{F}(U)$,

b) a homomorphism over R

$$A \to B .$$

This $\mathcal{F}$ is quasi-coherent if and only if:

$$B \cong A \otimes_R K .$$

THEOREM 1: If X is affine, and $\mathcal{F}$ is quasi-coherent, then
(A) $\mathcal{F}$ is spanned, as $\underline{o}_X$-module, by its sections $\Gamma(X, \mathcal{F})$,
(B) $H^i(X, \mathcal{F}) = (0)$, if $i > 0$.

We can now generalize these concepts in various ways:

Definition: Let X be a scheme. A sheaf $\mathcal{F}$ of $\underline{o}_X$-modules is quasi-coherent, if equivalently:

i) there exists a covering $\{U_i\}$ of X by affine open sets, such that $\mathcal{F}|_{U_i}$ is quasi-coherent;

ii) $\forall U \subset X$, U affine and open, $\mathcal{F}|_U$ is quasi-coherent.

A very useful application of this concept is in:

Proposition-Definition: Let X be a scheme. A closed sub-scheme $Y \subset X$ is a local ringed space Y whose underlying topological space is a closed subspace of X, and whose sheaf of rings $\underline{o}_Y$ is a quotient of $\underline{o}_X$: i.e., one has $0 \to \mathcal{I} \to \underline{o}_X \to \underline{o}_Y \to 0$ ($\mathcal{I}$ a sheaf of ideals in $\underline{o}_X$), provided that equivalently $\mathcal{I}$ is quasi-coherent, or Y is itself a scheme.

The fact that if Y is a scheme, then $\mathcal{I}$ is quasi-coherent comes from:

Proposition 2: Let $X \xrightarrow{f} Y$ be a quasi-compact morphism of schemes (i.e., if $U \subset Y$ is open and affine, $f^{-1}(U)$ admits a finite affine open covering). Then if $\mathcal{F}$ is a quasi-coherent sheaf on X, all the sheaves $R^i f_*(\mathcal{F})$ are quasi-coherent on Y.

One finds, from the above definition: the closed subschemes of $X = \text{Spec}(R)$ are the schemes $Y = \text{Spec}(R/I)$, for ideals $I \subset R$. We also make the definition:

Definition: If $Y \xrightarrow{f} X$ is an isomorphism of Y with a closed subscheme of X, then f is a closed immersion.

Definition: Let X be a scheme. A sub-scheme $Y \subset X$ is a closed subscheme of an open subset $U \subset X$. An immersion $Y \xrightarrow{f} X$ is an isomorphism of Y with a subscheme of X.

Example: One of the most important subschemes of a scheme X is X_{red} ("X reduced"). As a closed subset, $X_{red} = X$, but its defining sheaf of ideals $\mathcal{I}$ is the subsheaf:

$$\Gamma(U, \mathcal{I}) = \{s \in \Gamma(U, \underline{o}_X) \mid \text{Equivalently, } s(x) = 0, \text{ all } x \in U; \; s_x \in \underline{o}_x \text{ is nilpotent, all } x \in U\}$$

One checks that if $U = \text{Spec}(R)$, then $\mathcal{I}|_U$ is the sheaf $\tilde{I}$, where

$$I = \{a \in R \mid \text{Equivalently, } a \in \text{ every prime ideal } \wp; \text{ or } a \text{ is nilpotent}\}.$$

Therefore, $\mathcal{I}$ is quasi-coherent. (Compare Lecture 3, 1°).

Another generalization of the concept of "affine" is:

Definition: A morphism $X \xrightarrow{f} Y$ is affine if equivalently:

i) there exists an affine open covering $\{U_i\}$ of Y such that $f^{-1}(U_i)$ is affine, for all i;

ii) $\forall$ affine open sets $V \subset Y$, $f^{-1}(V)$ is affine.

Corollary of Theorem 1: If $X \xrightarrow{f} Y$ is affine, and the sheaf $\mathcal{F}$ of $\underline{o}_X$-modules is quasi-coherent, then:

(A) the canonical homomorphism:

$$f^*(f_* \mathcal{F}) \to \mathcal{F}$$

is surjective;

(B) $R^i f_*(\mathcal{F}) = (0)$, for $i > 0$.

The concepts of fibre product and affine morphisms are connected by the very simple but important:

Proposition 3: Let $X \xrightarrow{f} Y$ be an affine morphism, and let $Y' \xrightarrow{g} Y$ be an arbitrary morphism. We write X' for $X \times_Y Y'$ with morphisms labelled as follows:

$$\begin{array}{ccc} X' & \xrightarrow{g'} & X \\ {\scriptstyle f'}\downarrow & & \downarrow{\scriptstyle f} \\ Y' & \xrightarrow[g]{} & Y \end{array}$$

Then f' is an affine morphism. And if F is a quasi-coherent sheaf on X,

$$g^* f_*(\mathcal{F}) \underset{\text{(canonically)}}{\cong} f'_* g'^*(\mathcal{F}) .$$

2° We define several concepts by specializing the above to a more finite situation:

Definition: A scheme X is noetherian if, equivalently:

i) there exists a finite open affine covering $\{U_i\}$ of X such that $\Gamma(U_i, \underline{o}_X)$ is noetherian;

ii) X is quasi-compact, and for all open affine $U \subset X$, $\Gamma(U, \underline{o}_X)$ is noetherian;

iii) the ordered set of closed subschemes of X satisfies the descending chain condition.

Definition: A quasi-coherent sheaf $\mathcal{F}$ on a noetherian scheme X is coherent if, equivalently:

i) there exists an affine open covering $\{U_i\}$ of X such that $\Gamma(U_i, \mathcal{F})$ is a $\Gamma(U_i, \underline{o}_X)$-module of finite type;

ii) same for all affine open $U \subset X$.

Note. Quasi-coherent subsheaves and quotient sheaves of coherent sheaves are coherent; $\underline{o}_X$ is coherent; if the stalk $\mathcal{F}_x$ of a coherent sheaf $\mathcal{F}$ at x is (0), then $\mathcal{F} \cong (0)$ in a neighborhood of x.

Definition: An affine morphism $X \xrightarrow{f} Y$, where Y is noetherian, is finite if equivalently:

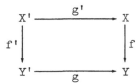

i) $f_*(\underline{o}_X)$ is coherent on Y;

ii) f is of finite type (hence X is noetherian) and for all coherent $\mathcal{F}$ on X, $f_*(\mathcal{F})$ is coherent on Y.

<u>Proposition 4</u>: If $X \xrightarrow{f} Y$ is finite, then for all $y \in Y$, the set of points $f^{-1}(y)$ is finite, (this property is what Grothendieck calls "quasi-finite").

<u>Proof:</u> If $A = f_*(\underline{o}_X)_y \otimes_{\underline{o}_y} \mathcal{K}(y)$, then it is easily seen that the scheme-theoretic fibre $f^{-1}(y)$ is simply Spec (A). But since $f_*(\underline{o}_X)$ is coherent, A is a finite dimensional $\mathcal{K}(y)$-algebra, hence Spec (A) is finite.

QED

Concerning the topology of noetherian schemes, the key point is that these are noetherian topological spaces, i.e., satisfy the d.c.c. for closed subsets. Consequently, every closed subset is a finite union of irreducible closed subsets which are called its <u>components</u>. This is, of course, the global topological analog of the decomposition of an ideal in a noetherian ring into an intersection of primary ideals. The finer aspects of the decomposition theorem come in via the operation "A":

Definition: Let $\mathcal{F}$ be a coherent sheaf on a noetherian scheme X.

$A(\mathcal{F}) = \{x \in X \mid \exists$ a section $s \in \mathcal{F}_x$ which is annihilated by an ideal $I \subset \underline{o}_x$ primary to the maximal ideal, i.e., $\exists$ an open neighborhood U of x, and $s \in \Gamma(U)$ such that the support of s is the closure of $x\}$

[cf. BOURBAKI, <u>Alg. Comm.</u>, Ch. 4, for a thorough discussion of this concept]. It follows immediately from the decomposition theorem for modules that $A(\mathcal{F})$ is a finite set. Moreover, $A(\mathcal{F})$ includes in particular, the generic points of every component of the support of $\mathcal{F}$ (as a closed subset of X)—but, in general, it also includes "embedded associated points." On the other hand, if Z is a closed subset of X and we make Z into a closed subscheme via the sheaf of <u>all</u> functions which are everywhere 0 on Z (this is known as the <u>reduced subscheme structure</u> on Z), then $A(\underline{o}_Z)$ is precisely the set of generic points of the components of Z.

3° Flatness:

<u>Definition:</u> Let $X \xrightarrow{f} Y$ be a morphism of schemes, and let $\mathcal{F}$ be a sheaf of $\underline{o}_X$-modules. Then $\mathcal{F}$ is <u>flat</u> over Y if for all $x \in X$, $\mathcal{F}_x$ is a flat $\underline{o}_{f(x)}$-module; $\mathcal{F}$ is of <u>finite Tor-dimension</u> over Y if there is an n such that for all $x \in X$, $\mathcal{F}_x$ is an $\underline{o}_{f(x)}$-module of Tor-dimension $\leq n$.

Using the fact that forming Tor's commutes with localization, one checks easily that, if $\mathcal{F}$ is quasi-coherent, then

(*) $\mathcal{F}$ is flat (resp. of f. Tor-dim.) over Y, if and only if for all affine open sets $V \subset Y$, $U \subset f^{-1}(V)$, the module $\Gamma(U, \mathcal{F})$ is flat (resp. of f. Tor-dim.) over the ring $\Gamma(V, \underline{o}_Y)$ (the Tor-dim. being bounded independently of U and V).

The key point of flatness is that it commutes with all base extensions: i.e., suppose

$$\begin{array}{ccc} X & \xleftarrow{\quad g' \quad} & X' \\ {\scriptstyle f}\downarrow & & \downarrow{\scriptstyle f'} \\ Y & \xleftarrow[\quad g \quad]{} & Y' \end{array}$$

given, where $X' \cong X \times_Y Y'$. Then if $\mathcal{F}$ is a sheaf of $\underline{o}_X$-modules, flat over Y, it follows immediately that the sheaf $g'^*(\mathcal{F})$ of $\underline{o}_{X'}$-modules is flat over Y'.

A priori, flatness would appear to be a fairly ungeometric concept. However, I think that this is untrue. The heuristic meaning of "$\mathcal{F}$ flat/Y" is that $\mathcal{F}$ induces a continuously varying family of sheaves on the fibres X_y of f. I think this is best shown by a series of illustrative examples:

<u>Example 1</u>: Assume that X and Y are noetherian and that $\mathcal{F}$ is coherent on X.

Now if $\mathcal{F}$ is to induce a continuously varying family of sheaves on X_y, surely a point of X at which $\mathcal{F}$ is exceptional should lie over a point of Y which is exceptional. In fact, one has:

<u>Proposition 5</u>: If $\mathcal{F}$ is flat/Y, and $x \in A(\mathcal{F})$, then $f(x) \in A(\underline{o}_Y)$.

<u>Proof</u>: Let $y = f(x)$. Recall that $y \in A(\underline{o}_Y)$ if and only if

(*) depth $(\underline{o}_y) = 0$, i.e., all non-units in $\underline{o}_y$ are 0-divisors.

Therefore, if $f(x) \notin A(\underline{o}_Y)$, there is a non-unit $a \in \underline{o}_y$ such that

$$0 \to \underline{o}_y \xrightarrow{a} \underline{o}_y$$

is injective. If $\mathcal{F}$ is flat /Y, it follows that:

$$0 \to \mathcal{F}_x \xrightarrow{f^*(a)} \mathcal{F}_x$$

is injective, where $f^*(a)$ is the induced non-unit of $\underline{o}_x$. But then multiplication by $f^*(a)^n$ is injective in $\mathcal{F}_x$, for all n, hence no $s \in \mathcal{F}_x$ is killed by an ideal primary to m_x.

A more precise result can be found in BOURBAKI: Alg. Comm. Ch.4, § . Namely, if $\mathcal{F}$ is flat /Y, then

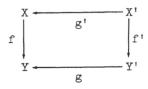

$$x \in A(\mathcal{F}) \iff \begin{array}{ll} \text{i)} & f(x) \in A(Y) \\ \text{ii)} & x \in A(\mathcal{F} \otimes \mathcal{K}(y)) \\ & \text{where } y = f(x). \end{array}$$

In fact, there is an even stronger result making use of the concept of depth: recall

Definition: Let $\mathcal{O}$ be a noetherian local ring, and let M be an $\mathcal{O}$-module of finite type. Then $d = \text{depth}(M)$ if there are exactly d elements in every maximal M-sequence $f_1, \ldots, f_d$ [i.e., in every sequence $f_1, \ldots, f_d \in m$ such that:

$$f_{i+1} \cdot a \in (f_1, \ldots, f_i) \cdot M \implies a \in (f_1, \ldots, f_i) \cdot M] .$$

Incidentally, one should regard the depth of $\mathcal{O}$ itself, for example, as a measure of the topological complexity of the singularity at the closed point of Spec ($\mathcal{O}$): if the depth is maximal, i.e., equals the dimension of $\mathcal{O}$, then $\mathcal{O}$ is, in a weak sense, non-singular, while if the depth is much less than the dimension, the singularity is very bad. The result a propos of flatness is:

THEOREM: For all $x \in X$, if $\mathcal{F}_x$ is flat over $\underline{o}_y$, $y = f(x)$, then

$$\text{depth}(\mathcal{F}_x) = \text{depth}(\underline{o}_y) + \text{depth}(\mathcal{F}_x) \otimes \mathcal{K}(y))$$

(the last being an $\underline{o}_x/m_y \cdot \underline{o}_x$-module). [Cf. EGA, 4, 6.3]

Example 2: Suppose we assume, in addition, that Y is a "non-singular curve," i.e., for all $y \in Y$, $\underline{o}_y$ is a regular local ring of dimension 0 or 1. Then we have converse:

Proposition 6:

$$[\mathcal{F} \text{ flat } /Y] \iff \left[\begin{array}{l} \text{for all } x \in A(\mathcal{F}), \ f(x) \text{ is a point} \\ \text{of } Y \text{ where } \underline{o}_y \text{ has dimension } 0 \end{array}\right]$$

Proof: We have proven "$\implies$". Now suppose $\mathcal{F}$ is not flat $/Y$, i.e., for some $x \in X$, $\mathcal{F}_x$ is not flat over $\underline{o}_y$, $y = f(x)$. Then $\underline{o}_y$ must have dimension 1: let $(\pi) \subset \underline{o}_y$ be its maximal ideal. But $\mathcal{F}_x$ is flat $/\underline{o}_y$ if and only if multiplication by $f^*(\pi)$ is injective in $\mathcal{F}_x$. Therefore, there is an $s \in \mathcal{F}_x$ such that $f^*(\pi) \cdot s = 0$. Let

$$\mathfrak{A} = \{t \in \underline{o}_x \mid t \cdot s = 0\} ,$$

and let $\wp$ be a prime ideal in $\underline{o}_x$, minimal among the prime ideals containing $\mathfrak{A}$. By Proposition 1, Lecture 3, there is a unique point $x' \in X$ such that x is in the closure of x', and $\underline{o}_{x'} = (\underline{o}_x)_\wp$. With respect to the given homomorphism:

$$\underline{o}_y \xrightarrow{f^*} \underline{o}_x \longrightarrow \underline{o}_{x'} ,$$

since $f^*(\pi) \in \mathfrak{A} \subset \wp$, the inverse image of the maximal ideal $m_{x'}$ is exactly m_y. By the remark following Theorem 1, Lecture 3, this means that $f(x') = y$. [i.e., use the diagram:

$$\begin{array}{ccc} \operatorname{Spec}(\underline{o}_x) & \hookrightarrow & X \\ \downarrow & & \downarrow \\ \operatorname{Spec}(\underline{o}_y) & \hookrightarrow & Y \end{array} \quad].$$

The proposition will therefore be proven if we verify that $x' \in A(\mathcal{F})$. But $\mathfrak{A}\underline{o}_{x'}$ is primary for the maximal ideal $m_{x'} \subset \underline{o}_{x'}$, and it kills the induced section $s' \in \mathcal{F}_{x'}$.

QED

__Example 3__: Now consider the case of a finite morphism $X \xrightarrow{f} Y$, Y noetherian, and a coherent sheaf $\mathcal{F}$ on X. The continuity of $\mathcal{F}$ over Y expresses itself as follows:

__Proposition 7__:

$$[\ \mathcal{F} \text{ flat } /Y] \iff [f_*\mathcal{F} \text{ is locally free on } Y]\ .$$

__Proof__: The result being local on Y, suppose $Y = \operatorname{Spec}(B)$; then $X = \operatorname{Spec}(A)$, where A is a B-algebra, and is of finite type as B-module. Let $\mathcal{F}$ correspond to the finite A-module M. If $\mathcal{F}$ is flat $/Y$, then M is flat $/B$, hence for all prime ideals $\wp \subset B$, $M_\wp = M \otimes_B B_\wp$ is flat over $B_\wp$, i.e., $f_*(\mathcal{F})_y = M_\wp$ is flat over $\underline{o}_y = B_\wp$. But a module of finite type over a noetherian local ring is flat only if it is free. Therefore, there is an isomorphism

$$\underline{o}_y^n \xrightarrow{\sim} f_*(\mathcal{F})_y$$

of $\underline{o}_y$-modules. But such a homomorphism is induced by a homomorphism:

$$\underline{o}_Y^n \longrightarrow f_*(\mathcal{F})$$

in some neighborhood of y; and the kernel and cokernel, having 0 stalks at y, also vanish in a neighborhood of y. Therefore $f_*(\mathcal{F})$ is locally free.

The converse is clear, since the stalk $\mathcal{F}_x$ at $x \in X$ is a localization of the $\underline{o}_{f(x)}$-module $f_*(\mathcal{F})_{f(x)}$.

QED

__Example 4__: We shall further analyze the situation of Example 3, in case Y is reduced and irreducible. Suppose $y \in Y$. Via the fibre of f over y, one has the diagram:

$$\begin{array}{ccc} X & \longleftarrow & X_y \\ \downarrow & & \downarrow \\ Y & \longleftarrow & \operatorname{Spec} \mathcal{K}(y) \end{array}$$

and $\mathcal{F}$ on X induces a sheaf $\mathcal{F}_y$ on X_y. Algebraically, if $Y = \text{Spec } B$, $X = \text{Spec }(A)$, and $\mathcal{F}$ corresponds to the A-module M, then y comes from a prime ideal $\wp \subset B$, $\kappa(y)$ is the quotient field of $B/\wp$,

$$X_y = \text{Spec }(A \underset{B}{\otimes} \kappa(y))$$

$$\mathcal{F}_y = \widetilde{M \underset{A}{\otimes} (A \underset{B}{\otimes} \kappa(y))} = \widetilde{M \underset{B}{\otimes} \kappa(y)} .$$

Since A is a finite B-module, $A \otimes_B \kappa(y)$ is a finite dimensional commutative algebra over $\kappa(y)$.

Note first of all that

$$(*) \qquad \Gamma(X_y, \mathcal{F}_y) \cong f_*(\mathcal{F}) \underset{\underline{o}_Y}{\otimes} \kappa(y) \cong M \underset{B}{\otimes} \kappa(y) .$$

(Cf. Proposition 3 of this lecture.)

Proposition 8:

$$[\ \mathcal{F} \text{ flat } /Y] \iff [\text{the function } y \to \dim_{\kappa(y)} f_*(\mathcal{F}) \otimes \kappa(y) \text{ is } \underline{\text{constant}}] .$$

Proof: The " $\Longrightarrow$" follows from Proposition 7, Y being irreducible and hence connected. To prove "$\Longleftarrow$", it suffices to show that for all $y \in Y$, $f_*(\mathcal{F})_y$ is a free $\underline{o}_y$-module.

Lemma: Let A be a noetherian local domain with residue field k, and quotient field K. Let M be a finite A-module. Then

$$[\dim_K M \underset{A}{\otimes} K = \dim_k M \underset{A}{\otimes} k] \Longrightarrow [M \text{ a free A-module}] .$$

Proof: Note that if $m \subset A$ is the maximal ideal, $M \otimes_A k \cong M/m \cdot M$. Let $f_1, \ldots, f_n$ be elements of M whose images $\overline{f_i}$ in $M/m \cdot M$ form a basis over k. Then the f_i define a homomorphism φ:

$$(*) \qquad 0 \to L \to A^n \xrightarrow{\varphi} M \to N \to 0$$

(L and N being the kernel and cokernel resp.). Tensoring with k, we obtain:

$$k^n \xrightarrow{\overline{\varphi}} M/m \cdot M \to N/m \cdot N \to 0 .$$

But $\overline{\varphi}$ is surjective since the $\overline{f_i}$ span $M/m \cdot N$; therefore, $N = m \cdot N$. By Nakayama's lemma, $N = (0)$. Now tensor (*) with K. Since K is flat $/A$, we obtain:

$$0 \to L \underset{A}{\otimes} K \to K^n \to M \underset{A}{\otimes} K \to 0 .$$

By hypothesis, K^n and $M \otimes_A K$ are both K-vector spaces of dimension n. Therefore, $L \otimes_A K = (0)$, i.e., L is a torsion module. But since $L \subset A^n$, this implies that $L = (0)$. QED

Example 5: As a final point, let us consider two completely concrete cases:

(I) $Y = \text{Spec } k[y]$
$X = \text{Spec } k[x]$
$y = x^2$.
(k alg. closed).

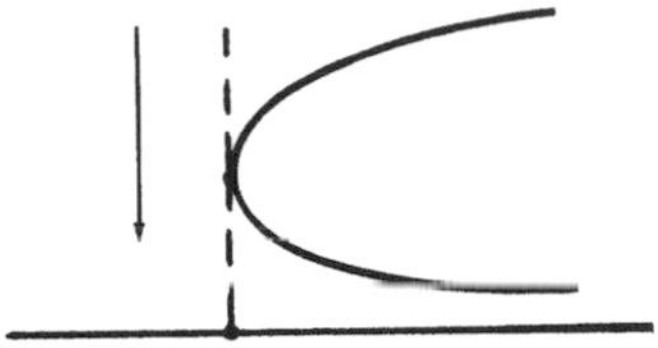

Then if $\wp \subset k[y]$ is the maximal ideal $(y - \alpha^2)$,

$$k[x]/\wp \cdot k[x] \cong k[x]/(x - \alpha) \oplus k[x]/(x + \alpha), \quad \alpha \neq 0$$

$$k[x]/\wp \cdot k[x] \cong k[x]/(x^2), \quad \alpha = 0,$$

and both are commutative algebras of dimension 2 over k. This being a constant f is flat. (One should also check non-closed points of Y.)

(II) $Y = \text{Spec } k[x_1^2, x_1x_2, x_2^2]$
$X = \text{Spec } k[x_1, x_2]$

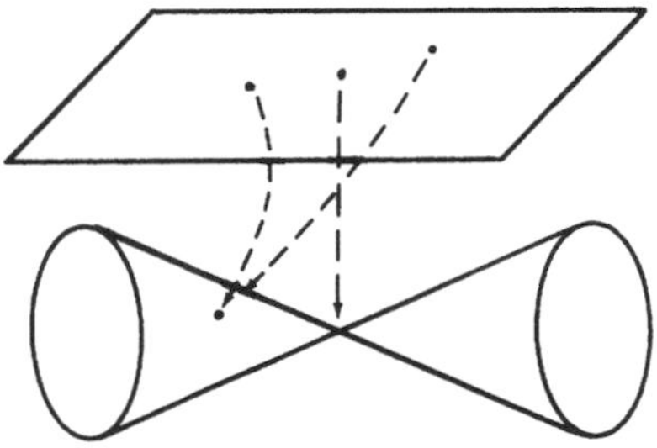

Then if $\wp \subset k[x_1^2, x_1x_2, x_2^2]$ is the maximal ideal $(x_1^2 - \alpha^2, x_1x_2 - \alpha\beta, x_2^2 - \beta^2)$, one finds

$$\begin{aligned} k[x_1, x_2]/\wp \cdot k[x_1, x_2] &\cong k[x_1, x_2]/(x_1 - \alpha, x_2 - \beta) \\ &\oplus k[x_1, x_2]/(x_1 + \alpha, x_2 + \beta) \\ &\text{if } \alpha \text{ or } \beta \neq 0, \end{aligned}$$

$$\begin{aligned} k[x_1, x_2]/\wp \cdot k[x_1, x_2] &\cong k + x_1 \cdot k + x_2 \cdot k \\ &(x_1^2 = x_1x_2 = x_2^2 = 0) \\ &\text{if } \alpha = \beta = 0 . \end{aligned}$$

The former is a commutative algebra of dimension 2; the latter is one of dimension 3. Therefore f is not flat.

LECTURE 7

RESUME OF THE COHOMOLOGY OF COHERENT SHEAVES ON $\mathbf{P}_n$

As above, let $\mathbf{P}_n = \text{Proj } \mathbf{Z}[X_0,\ldots, X_n]$, let $\underline{o}(1)$ be the canonical sheaf on $\mathbf{P}_n$, and identify $X_0,\ldots, X_n$ with sections of $\underline{o}(1)$. For all schemes S, on $\mathbf{P}_n \times S$, put

$$\underline{o}(1) = p_1^*(\underline{o}(1)) \quad \text{(by abuse of language)}$$

$$X_i = \text{the induced section } p_1^*(X_i) \quad \text{(by abuse of language)}.$$

If $\mathcal{F}$ is a coherent sheaf on $\mathbf{P}_n \times S$, put

$$\mathcal{F}(m) = \mathcal{F} \underset{\underline{o}_{\mathbf{P}_n \times S}}{\otimes} (\underline{o}(1)^{\otimes m}) .$$

1° Serre's results. We look first at the readily visualized case $S = \text{Spec }(k)$, k a field. Fix $\mathcal{F}$ (again coherent), and write $\mathbf{P}_{n,k}$ for $\mathbf{P}_n \times \text{Spec }(k)$:

(i) $H^i(\mathbf{P}_{n,k}, \mathcal{F})$ is finite dimensional over k, for all i; and is (0) if $i > n$;

(ii) For all $\mathcal{F}$, there exists m_0 such that if $m \geq m_0$ $H^i(\mathbf{P}_{n,k}, \mathcal{F}(m)) = (0)$, $i > 0$ and $\mathcal{F}(m)$ spanned, as $\underline{o}_{\mathbf{P}_{n,k}}$-module, by its global sections;

(iii) $\Sigma_{i=0}^n (-1)^i \dim_k H^i(\mathbf{P}_{n,k}, \underline{\mathcal{F}}(m))$ is a polynomial in m — the Hilbert polynomial of $\mathcal{F}$.

(iv) Consider the functor:

$$\alpha : \mathcal{F} \to \bigoplus_{m=0}^{\infty} \Gamma(\mathbf{P}_{n,k}, \mathcal{F}(m)) .$$

Here $\mathcal{F}$ is an object in the category $\mathcal{C}$ of coherent sheaves on $\mathbf{P}_{n,k}$; and $\alpha(\mathcal{F})$ is an object in the category $\mathcal{C}'$ of graded $k[X_0,\ldots, X_n]$-modules of finite type.

[If $t \in \Gamma(\mathbf{P}_{n,k}, \mathcal{F}(m))$, then $X_i \cdot t$ is the section $t \otimes X_i$ of $\mathcal{F}(m) \otimes \underline{o}(1) \cong \mathcal{F}(m+1)$.

Take morphisms in $\mathcal{C}'$ to be:

$$\mathrm{Hom}_{\mathcal{C}'}(M, N) = \varinjlim_{m_0} \mathrm{Hom}_{\substack{\text{Gradation}\\ \text{preserving}}} \left[\bigoplus_{m \geq m_0} M_m, \bigoplus_{m \geq m_0} N_m \right] .$$

Then α is an equivalence of categories, especially α is exact, and takes Hom's into Hom's. The key step in proving this is the explicit construction of the inverse of α. This functor is a graded generalization of the $\sim$ operation in the affine case. Start with a graded module M, of finite type over $k[X_0, \ldots, X_n]$. For each i, form the tensor product.

$$M^{(i)} = M \underset{k[X]}{\otimes} k[X_0, \ldots, X_n, \frac{1}{X_i}] ,$$

and let $M_0^{(i)}$ be the sub-module of degree 0. Then $M_0^{(i)}$ is a module of finite type over the affine coordinate ring $k[X_0/X_i, \ldots, X_n/X_i]$ of $(\mathbf{P}_n)_{X_i}$. One verifies that the sheaves $M_0^{(i)}$ on the affine spaces patch together in a natural way: the result is called $\widetilde{M}$ and this is the inverse of α.

(v) Before proceeding to generalizations, we want to make some attempt to describe the "yoga" of cohomology. The cohomology of sheaves, in a general geometric setting, is just a piece of machinery designed to analyze the connection between the local and global structure of space; viz. given any local data, the set of all such local data will form a sheaf and its cohomology groups are a sequence of invariants describing how "twisted" these data can be from a global point of view. The essential point is that (a) these groups are almost always very computable, (b) the obstructions to making global constructions are elements of such cohomology groups.

In the case of algebraic geometry, the objects of global geometric interest are the global sections of coherent sheaves. These arise for example out of the desire to determine how many functions exist on some scheme with prescribed poles; in what projective spaces can a given scheme be embedded; how many global differential forms of a given type exist on some scheme; and in the infinitesimal linear form of many non-linear existence problems. But to compute the vector space of sections of a coherent sheaf $\mathcal{F}$ on $\mathbf{P}_n$, the essential difficulty is that Γ is not a right exact functor. This was realized by the Italian geometers, who worked indirectly but still (as we now realize) very closely with the higher cohomology groups.

It should be pointed out that the fancy definitions given cohomology recently—via standard resolutions, derived functors, especially in the category of all sheaves—which look very uncomputable—are just technical devices to simplify somebody's general theory. One may as well treat the cohomology of a coherent sheaf on $\mathbf{P}_n$ just as the satellites of Γ in the workable category of coherent sheaves. [In technical terms, cohomology is <u>effacable</u> in this small category]: e.g., the group $H^1(\mathbf{P}_1, \underline{o}_{\mathbf{P}_1}(-2)) \cong k$ is nothing but the cokernel of the sequence:

$$0 \to \Gamma(\mathbf{P}_1, \underline{o}_{\mathbf{P}_1}(-2)) \to \Gamma(\mathbf{P}_1, \underline{o}_{\mathbf{P}_1}(-1)) \to \Gamma(\mathbf{P}_1, \mathcal{H}(x))$$

coming from the exact sequence of sheaves:

$$0 \to \underline{o}_{\mathbf{P}_1}(-2) \xrightarrow{\otimes X_1} \underline{o}_{\mathbf{P}_1}(-1) \to \mathcal{H}(x) \to 0$$

on $\mathbf{P}_1$, where $\mathcal{H}(x)$ is the sheaf with support only at the point x where $X_1 = 0$, given by the module which is the residue class field of $\underline{o}_x$.

We must recall, for future use, the facts about the cohomology of $\underline{o}_{\mathbf{P}_n}(m)$ itself:

$$\begin{aligned} H^i(\mathbf{P}_{n,k}, \underline{o}_{\mathbf{P}_n}(m)) &= (0), \text{ if } 0 < i < n \\ &= (0), \text{ if } i = n,\ m > -n-1 \\ &= (0), \text{ if } i = 0,\ m < 0 \\ &= \left.\begin{array}{l}\text{a vector space with basis given}\\ \text{by the monomials in } X_0, \ldots, X_n\\ \text{of degree } m.\end{array}\right\} \begin{array}{l} i = 0 \\ m \geq 0\end{array} \end{aligned}$$

2° <u>Grothendieck's globalization.</u> Now suppose S is any noetherian scheme, and $\mathcal{F}$ is again coherent on $\mathbf{P}_n \times S$. Let $p\colon \mathbf{P}_n \times S \to S$ be the projection. Then:

(i) $R^i p_*(\mathcal{F})$ is coherent for all i; and is (0) if $i > n$.

(ii) For all $\mathcal{F}$, there exists m_0 such that if $m \geq m_0$,
$R^i p_*(\mathcal{F}(m)) = (0)$, $i > 0$, and
$p^* p_* \mathcal{F}(m) \to \mathcal{F}(m)$ is surjective.

(iii) Consider the functor:

$$\alpha\colon \mathcal{F} \to \bigoplus_{m=0}^{\infty} p_*(\mathcal{F}(m)) .$$

Here $\mathcal{F}$ is an object in the category $\mathcal{C}$ of coherent sheaves of $\underline{o}_{\mathbf{P}_n \times S}$-modules; and $\alpha(\mathcal{F})$ is an object in the category $\mathcal{C}'$ of quasi-coherent sheaves of graded $\underline{o}_S[X_0, X_1, \ldots, X_n]$-modules of finite type—where the morphisms are given by:

$$\operatorname{Hom}_{\mathcal{C}'}(\mathfrak{M}, \mathfrak{N}) = \varinjlim_{m_0} \operatorname{Hom}_{\substack{\text{Gradation}\\ \text{preserving}}} \left[\bigoplus_{m \geq m_0} \mathfrak{M}_m ; \bigoplus_{m \geq m_0} \mathfrak{N}_m \right].$$

Then α is an equivalence of categories.

In fact, the inverse $\sim$ to α is constructed exactly as in 1°: start with the sheaf $\mathfrak{M}$ on S. For simplicity, assume S is affine, say $S = \operatorname{Spec}(R)$. Then $\mathfrak{M}$ is nothing but a graded $R[X_0, \ldots, X_n]$-module of finite type. For all i, put

$$\mathfrak{M}_0^{(i)} = \text{degree 0 component of } \left(\mathfrak{M} \underset{R[X]}{\otimes} R[X_0, \ldots, X_n, \frac{1}{X_i}] \right)$$

Then $\widetilde{\mathfrak{M}}$ is patched together out of the sheaves $\widetilde{\mathfrak{M}}_0^{(i)}$ on:

$$\text{Spec } R \left[\frac{X_0}{X_i}, \ldots, \frac{X_n}{X_i} \right] = (\mathbf{P}_n \times S)_{X_i} .$$

3° <u>Connection of higher direct images with cohomology on the fibres</u>. The principle difficulty in using the results of 2° is in relating $R^i p_*(\mathfrak{F})$ to the cohomology along the fibres of p. Thus, if $s \in S$, let $\mathbf{P}_{n,s}$ = the fibre of p over s, and let $\mathfrak{F}$ induce the coherent sheaf $\mathfrak{F}_s$ on $\mathbf{P}_{n,s}$. Is there any connection between;

$$R^i p_*(\mathfrak{F}) \otimes \mathfrak{K}(s) \quad \underline{\text{and}} \quad H^i(\mathbf{P}_{n,s}, \bar{\mathfrak{F}}_s) .$$

This is a special case of the more general problem; given a "base extension" $g\colon T \to S$, look at the diagram:

$$\begin{array}{ccc} \mathbf{P}_n \times T & \xrightarrow{\ h\ } & \mathbf{P}_n \times S \\ {\scriptstyle q}\downarrow & & \downarrow{\scriptstyle p} \\ T & \xrightarrow{\ g\ } & S \end{array}$$

What is the relation between

$$g^* R^i p_*(\mathfrak{F}) \text{ and } R^i q_*(h^* \mathfrak{F}) ,$$

for coherent sheaves $\mathfrak{F}$ on $\mathbf{P}_n \times S$? But, for any open set $U \subset S$, one has homomorphisms:

$$H^i(\mathbf{P}_n \times U, \mathfrak{F}) \to H^i(\mathbf{P}_n \times g^{-1}(U), h^* \mathfrak{F}) \to H^0(g^{-1}(U), R^i q_*(h^* \mathfrak{F}))$$

hence a homomorphism:

$$R^i p_*(\mathfrak{F}) \to g_* \ R^i q_*(h^* \mathfrak{F})$$

hence a homomorphism: $g^* R^i p_*(\mathfrak{F}) \to R^i q_*(h^* \mathfrak{F})$.

If, for every g, this is an isomorphism, we shall say that $R^i p_*$ <u>commutes with base extension.</u>

First of all, there is a simple "stable" result when $\mathfrak{F}$ has been twisted sufficiently:

(i) For any $\mathfrak{F}$, and any $T \xrightarrow{g} S$, there is an m_0 such that if $m \geq m_0$, then:

$$g^* p_*(\mathfrak{F}(m)) \xrightarrow{\sim} q_* h^*(\mathfrak{F}(m))$$

(of course, both sets of <u>higher</u> direct images are zero).

Idea of proof: This really asserts nothing more than the compatibility of the equivalences of categories α_S and α_T with tensor products. Thus, over S, $\mathcal{F}$ is defined by the sheaf of graded $\underline{o}_S[X_0, \ldots, X_n]$-modules:

$$\alpha_S(\mathcal{F}) = \mathfrak{M} = \bigoplus_{m=0}^{\infty} p_*(\mathcal{F}(m))$$

and, over T, $h^* \mathcal{F}$ is defined by the sheaf of graded $\underline{o}_T[X_0, \ldots, X_n]$-modules:

$$\alpha_T(h^* \mathcal{F}) = \mathfrak{N} = \bigoplus_{m=0}^{\infty} q_*[h^*(\mathcal{F}(m))] .$$

One wants to know that the natural homomorphism from $g^* \mathfrak{M}$ to $\mathfrak{N}$ is an isomorphism in our funny category (where any finite number of graded pieces can be ignored). To prove this, use the inverse $\sim$ to α! Since α_S and α_T are equivalences of categories, it suffices to prove that

$$\widetilde{g^* \mathfrak{M}} \cong h^*(\widetilde{\mathfrak{M}}) .$$

But this is an immediate consequence of the definition of $\sim$ [for details, cf. EGA, Ch. 2, §§2.8.10 when S, T affine; 3.5.3 in general].

However, to obtain really precise relations between these higher direct images, we must look at the case when $\mathcal{F}$ is flat over S;

(ii) Assume $\mathcal{F}$ is flat over S, and that for some i, and some $s_0 \in S$, the homomorphism:

$$R^i p_*(\mathcal{F}) \otimes \mathcal{K}(s_0) \to H^i(\mathbf{P}_{n,s_0}, \mathcal{F}_{s_0})$$

is surjective. Then there is an open neighborhood U of s_0 in S such that for any base extension $g: T \to U$, the homomorphism

$$g^* R^i p_*(\mathcal{F}) \xrightarrow{\sim} R^i q_*(h^* \mathcal{F})$$

is an isomorphism. (See EGA, Ch. 3, §7.7.)

(iii) With the same assumptions as in (ii), it follows that the homomorphism:

$$R^{i-1} p_*(\mathcal{F}) \otimes \mathcal{K}(s_0) \to H^{i-1}(\mathbf{P}_{n,s_0}, \mathcal{F}_{s_0})$$

is also surjective if and only if $R^i p_*(\mathcal{F})$ is a free sheaf of $\underline{o}_S$-modules in some neighborhood of s_0. (See EGA, Ch. 3, § 7.8.)

Corollary 1: In the flat case, if $H^{j+1}(\mathbf{P}_{n,s_0}, \mathcal{F}_{s_0}) = (0)$, then there is an open $U \subset S$ containing s_0 such that, for $g: T \to U$:

$$g^* R^j p_*(\mathcal{F}) \xrightarrow{\sim} R^j q_*(h^* \mathcal{F}) .$$

In particular:

$$R^j p_*(\mathcal{F}) \otimes \mathcal{K}(s) \xrightarrow{\sim} H^j(\mathbf{P}_{n,s}, \mathcal{F}_s) ,$$

for all $s \in U$.

Proof: Use (iii) for $i = j+1$ and then (ii) for $i = j$.

Corollary $1\frac{1}{2}$: In the flat case, if $R^i p_*(\mathcal{F}) = (0)$, for all $i \geq i_0$, then $H^i(\mathbf{P}_{n,s}, \mathcal{F}_s) = (0)$ for all $s \in S$, and all $i \geq i_0$.

Proof: Apply Corollary 1 first for $j = n$ to prove that $H^n(\mathbf{P}_{n,s}, \mathcal{F}_s) = (0)$, all $s \in S$; then for $j = n-1$ to prove that $H^{n-1}(\mathbf{P}_{n,s}, \mathcal{F}_s) = (0)$, all $s \in S$; etc.

Corollary 2: In the flat case, given a coherent sheaf $\mathcal{E}$ on S, and a homomorphism φ from $\mathcal{E}$ to $p_*(\mathcal{F})$ such that the induced

$$\mathcal{E} \otimes \mathcal{K}(s) \to H^0(\mathbf{P}_{n,s}, \mathcal{F}_s)$$

is an isomorphism for all s, then φ is an isomorphism, $\mathcal{E}$ is a locally free sheaf, and

$$g^* p_* \mathcal{F} \xrightarrow{\sim} q_* h^* \mathcal{F}$$

for all g.

Proof: Apply (ii) for $i = 0$, and (iii) for $i = 0$. Then use Nakayama's lemma.

Corollary 3: Given a coherent sheaf $\mathcal{F}$ on $\mathbf{P}_n \times S$, $\mathcal{F}$ is flat over S if and only if there exists an m_0 such that if $m \geq m_0$, $p_*(\mathcal{F}(m))$ is locally free. Hence, in this case, the Hilbert polynomial of $\mathcal{F}_s$ on $\mathbf{P}_{n,s}$ is locally constant.

Proof: If $\mathcal{F}$ is flat over S, then let m_0 be large enough so that $R^i p_*(\mathcal{F}(m)) = (0)$, if $i > 0$, $m \geq m_0$. Using Corollary 1 and $1\frac{1}{2}$ one deduces that $p_*(\mathcal{F}(m)) \otimes \mathcal{K}(s)$ maps onto $H^0(\mathbf{P}_{n,s}, \mathcal{F}_s(m))$ for all s, $m \geq m_0$. Then by (iii), $p_*(\mathcal{F}(m))$ is locally free. As for the converse, the point is that

$$\alpha(\mathcal{F}) = \bigoplus_{m=0}^{\infty} p_*(\mathcal{F}(m))$$

is a flat $\underline{o}_S$-module after throwing away a finite number of terms. Again using the $\sim$ operation inverse to α, it comes out immediately that $\mathcal{F}$ is defined over suitable affine sets by modules obtained in 2 steps:

(a) localizing $\alpha(\mathcal{F})$ with respect to X_i;

(b) passing to the sub-module of degree 0, which is a direct summand.

These are certainly flat over $\underline{o}_S$ if $\alpha(\mathcal{F})$ is flat, hence $\mathcal{F}$ is flat /S. (Cf. EGA, Ch. 3, §7.9.14.)

QED

Corollary 4: The projection $p: \mathbf{P}_n \times S \to S$ is (topologically) closed.

Proof: Let $Z \subset \mathbf{P}_n \times S$ be a closed subset. Let $\mathcal{F}$ be the structure sheaf of the reduced closed subscheme with support Z. By 2°, pick an m_0 such that $p^*p_*(\mathcal{F}(m)) \to \mathcal{F}(m)$ is surjective if $m \geq m_0$.

I claim:

$$p(Z) = \bigcap_{m \geq m_0} \text{Support } [p_*(\mathcal{F}(m))].$$

Since the sections of $p_*(\mathcal{F}(m))$ generate $\mathcal{F}(m)$, it follows first that $p_*(\mathcal{F}(m))_s \neq (0)$ for any $s \notin p(Z)$. Therefore $p(Z)$ is contained in the intersection. On the other hand, suppose $s \notin p(Z)$: then $\mathcal{F}_s = (0)$. By result (1) of 3°, for large enough m

$$p_*(\mathcal{F}(m)) \otimes \mathcal{K}(s) \xrightarrow{\sim} H^0(\mathbf{P}_{n,s}, \mathcal{F}(m)_s) = (0) ;$$

hence, by Nakayama's lemma $p_*(\mathcal{F}(m))_s = (0)$.

QED

<u>Corollary 5</u>: $R^i p_*(\underline{o}(m)) = (0)$, if $0 < i < n$
$= (0)$, if $i = n$, $m > -n-1$
$=$ free sheaf of $\underline{o}_S$-modules, with basis given by monomials in $X_0, \ldots, X_n$ of degree m, if $i = 0$.

<u>Proof</u>: Use 3° (ii) and (iii) and 2° (v). QED

4° It seems worthwhile to give one non-trivial example of this theory:

(i) Let $n = 1$, $S = \text{Spec } k[t]$, k an algebraically closed field

$$\mathbf{P}_1 \times S = \text{Proj } k[t; X_0, X_1]; \quad \text{let } R = k[t; X_0, X_1].$$

(ii) For all integers m, and graded R-modules M, put $M(m)$ equal to the R-module such that

$$M(m)_k = M_{m+k} \,.$$

(iii) Define the graded module M as

$$\left[R \oplus R \oplus R(-1) \big/ \begin{matrix}\text{modulo the element } (X_0, X_1, t) \\ \text{of degree } 1.\end{matrix} \right]$$

Put $\mathcal{F} = \widetilde{M}$. Corresponding to its definition as module, $\mathcal{F}$ is the cokernel in:

$$0 \rightarrow \underline{o}_{\mathbf{P}_1 \times S}(-1) \xrightarrow{\psi} \underline{o}_{\mathbf{P}_1 \times S} \oplus \underline{o}_{\mathbf{P}_1 \times S} \oplus \underline{o}_{\mathbf{P}_1 \times S}(-1) \rightarrow \mathcal{F} \rightarrow 0$$

where $\psi = (X_0, X_1, t)$ [i.e., tensoring with X_i maps $\underline{o}_{\mathbf{P}_1 \times S}(k)$ to $\underline{o}_{\mathbf{P}_1 \times S}(k+1)$; and multiplication by the ordinary function t maps $\underline{o}_{\mathbf{P}_1 \times S}(k)$ to $\underline{o}_{\mathbf{P}_1 \times S}(k)$.] Since the map ψ_x gotten by tensoring ψ with $\mathcal{K}(x)$, $(x \in \mathbf{P}_1 \times S)$, is never 0, it follows that $\mathcal{F}$ is a locally free sheaf of rank 2, and it is flat over S.

(iv) Let $0 \in S$ be the point $t = 0$. Then the induced sheaf $\mathcal{F}_0$ is defined by:

$$0 \to \underline{o}_{\mathbf{P}_1}(-1) \xrightarrow{(X_0,\, X_1,\, 0)} \underline{o}_{\mathbf{P}_1} \oplus \underline{o}_{\mathbf{P}_1} \oplus \underline{o}_{\mathbf{P}_1}(-1) \to \mathcal{F}_0 \to 0 .$$

and one checks that this means:

$$\mathcal{F}_0 \cong \underline{o}_{\mathbf{P}_1}(+1) \oplus \underline{o}_{\mathbf{P}_1}(-1) \quad .$$

On the other hand, if $s \in S$ is a k-rational point where $t = \alpha \neq 0$, $(\alpha \in k)$, then the diagram:

$$\begin{array}{ccccccc}
0 \to \underline{o}_{\mathbf{P}_1}(-1) & \xrightarrow{\psi_s} & \underline{o}_{\mathbf{P}_1} \oplus \underline{o}_{\mathbf{P}_1} \oplus \underline{o}_{\mathbf{P}_1}(-1) & \to & \mathcal{F}_s & \to & 0 \\
\downarrow & & \downarrow \varphi_s & & \downarrow & & \\
0 & \longrightarrow & \underline{o}_{\mathbf{P}_1} \oplus \underline{o}_{\mathbf{P}_1} & \longrightarrow & \underline{o}_{\mathbf{P}_1} \oplus \underline{o}_{\mathbf{P}_1} & \to & 0
\end{array}$$

where φ_s is defined by the 2×3 matrix

$$\begin{pmatrix} 1 & 0 & -X_0/\alpha \\ 0 & 1 & -X_1/\alpha \end{pmatrix}$$

makes $\mathcal{F}_s$ isomorphic to $\underline{o}_{\mathbf{P}_1} \oplus \underline{o}_{\mathbf{P}_1}$.

(v) The cohomologically interesting point is:

$$\begin{cases} p_*(\mathcal{F}(-1)) = (0) \\ H^0(\mathbf{P}_{1,0}, \mathcal{F}_0(-1)) \cong k , \end{cases}$$

i.e., p_* does not map onto the H^0 along the fibre; which is consistent with the theory in view of:

$$\begin{cases} H^1(\mathbf{P}_{1,0}, \mathcal{F}_0(-1)) \cong k \\ R^1 p_*(\mathcal{F}(-1)) \cong k_0 , \end{cases}$$

i.e., the sheaf concentrated at $t = 0$, which as module is the residue class field k of $\underline{o}_{0,S}$.

[Prove this by setting up an exact sequence

$$0 \to \mathcal{F}(-1) \to \mathcal{F} \to \underline{o}_Z \oplus \underline{o}_Z \to 0$$

where $Z \subset \mathbf{P}_1 \times S$ is the closed subscheme $X_1 = 0$, using the results of 3° to compute $R^i p_*(\mathcal{F})$, and using the cohomology sequence.]

LECTURE 8

FLATTENING STRATIFICATIONS

The problem we want to consider is this: Given a coherent sheaf $\mathcal{F}$ on $P_n \times S$, S a noetherian scheme—for all morphisms $T \xrightarrow{g} S$, one has the induced sheaf:

$$\mathcal{F}_g = (1_P \times g)^* \mathcal{F} \quad \text{on } P_n \times T .$$

Can you describe the set of all morphisms g such that $\mathcal{F}_g$ is flat over T ? To answer this, we first make:

<u>Definition</u>: If S is a scheme, a <u>stratification</u> of S is a finite set $S_1, \ldots, S_m$ of locally closed subschemes of S such that every point $s \in S$ is in exactly one subset S_i.

THEOREM: In the above situation, there is a stratification $S_1, \ldots, S_m$ of S such that for all morphisms $T \xrightarrow{g} S$ (T <u>noetherian</u>), $\mathcal{F}_g$ is flat over T if and only if the morphism g factors:

$$T \xrightarrow{g'} \coprod_{i=1}^{m} S_i \hookrightarrow S .$$

We will call this a flattening stratification: If it exists, it is obviously unique. There is an analogous problem when $P_n \times S$ is replaced by any scheme X proper over S. Grothendieck has then proven a slightly weaker theorem, but by <u>much</u> deeper methods.

1° Look first at the case $n = 0$; $\mathcal{F}$ is a coherent sheaf on S itself. Now $\mathcal{F}_g$ is simply $g^*(\mathcal{F})$, and it is flat over T if and only if it is locally free over T. For all $s \in S$, let

$$e(s) = \dim_{\mathcal{K}(s)} \left(\mathcal{F}_s \otimes_{\underline{O}_s} \mathcal{K}(s) \right) .$$

Fix a point s for a while, let $e = e(s)$, and choose $a_1, \ldots, a_e \in \mathcal{F}_s$ whose images in $\mathcal{F}_s \otimes \mathcal{K}(s)$ are a basis of this vector space. Then these a_i extend to sections of $\mathcal{F}$ in a neighborhood U_1 of s, and via the a_i one defines a homomorphism:

$$\underline{o}_S^e \xrightarrow{\varphi} \mathcal{F}$$

in U_1. Since the a_i generate $\mathcal{F}_s \otimes \mathcal{K}(s)$, by Nakayama's lemma, the a_i generate $\mathcal{F}_s$ itself. Therefore the homomorphism φ is surjective in a (possibly) smaller neighborhood U_2 of s. Passing to an even smaller neighborhood U_3, we may assume that Ker (φ) is generated by its sections over U_3, and we have constructed an exact sequence:

$$\underline{o}_S^f \xrightarrow{\psi} \underline{o}_S^e \xrightarrow{\varphi} \mathcal{F} \to 0$$

in U_3 (for some f). Let U_3 be called U_s.

Note first of all that $\mathcal{F}$ is generated by $e(s)$ sections everywhere in U_3, hence:

(*) $$\text{if } s' \in U_s, \quad e(s') \leq e(s) .$$

i.e., e is upper semi-continuous. Therefore the set

$$Z_e = \{s \in S \mid e(s) = e\}$$

is locally closed. Moreover, if $s' \in U_s$, then $e(s') = e(s)$ if and only if the homomorphism

$$\mathcal{K}(s')^f \xrightarrow{\psi(s')} \mathcal{K}(s')^e$$

is 0. Therefore, if ψ is expressed by an $e \times f$ matrix ψ_{ij} of functions on U_s, the closed subscheme Y_s of U_s defined by the ideal $(\psi_{ij})_{\text{all } i,j}$ has support $Z_e \cap U_s$. I claim that Y_s has the property:

(*) if $T \xrightarrow{g} U_s$ is any morphism (T noetherian), then $g^*(\mathcal{F})$ is locally free of rank $e = e(s)$ if and only if g factors through the closed subscheme Y_s.

Proof of *: g factors through Y_s if and only if all the functions $g^*(\psi_{ij})$ are 0 on T. But since the sequence:

$$\underline{o}_T^f \xrightarrow{g^*(\psi)} \underline{o}_T^e \xrightarrow{g^*(\varphi)} g^*(\mathcal{F}) \to 0$$

is exact on T, this is equivalent to asserting that $g^*(\varphi)$ is an isomorphism. Certainly this in turn implies that $g^*(\mathcal{F})$ is locally free of rank e; conversely, say $g^*(\mathcal{F})$ is locally free of rank e, and let $\mathcal{G}$ be the kernel of $g^*(\varphi)$. Tensoring with the residue field k at any point $t \in T$, one finds:

$$\begin{array}{c} \operatorname{Tor}_1(g^*\mathcal{F}, k) \to \mathcal{G} \otimes k \to k^e \to g^*(\mathcal{F}) \otimes k \to 0 \\ \| \\ (0) \end{array}$$

Since $g^*(\mathcal{F}) \otimes k$ is a k-vector space of dimension e, $\mathcal{G} \otimes k = (0)$, hence by Nakayama's lemma, $\mathcal{G} = (0)$ near t. Therefore $\mathcal{G} = (0)$ everywhere, and $g^*(\varphi)$ is an isomorphism.

QED

Note that property (*) characterizes the subscheme Y_s in a neighborhood of any point of $Z_e \cap U_s$. Therefore, if s_1 and s_2 are any two points of Z_e, in the open set $U_{s_1} \cap U_{s_2}$ the two subschemes Y_{s_1} and Y_{s_2} are equal. In other words, the subschemes Y_s patch together to endow the locally closed subset Z_e with a structure of subscheme. Call this subscheme Y_e. The collection $\{Y_e\}$ is a stratification of S, and, by virtue of (*), it follows immediately that $\{Y_e\}$ is a flattening stratification for $\mathcal{F}$.

For use in 3°, I want to make explicit that we have proven more than that a flattening stratification $\{Y_e\}$ exists: We have even indexed the subschemes Y_e so that $\mathcal{F} \otimes_{\underline{O}_S} \underline{O}_{Y_e}$ is locally free of rank e.

2° Before attacking the general case of the theorem, we need an elegant piece of "hard" algebra (cf. EGA, Ch. 4, §6.9) which gives us something to start with:

Proposition: Let $X \xrightarrow{f} Y$ be a morphism of finite type of noetherian schemes, and let $\mathcal{F}$ be a coherent sheaf on X. Assume that Y is reduced and irreducible. Then there is a non-empty open subset $U \subset Y$ such that the restriction of $\mathcal{F}$ to $h^{-1}(U)$ is flat over U.

Proof: We may clearly replace Y be some affine open subset Spec (A); and since X can be covered by a finite set of affine open subsets V_i, it clearly suffices to find one U for each V_i so that in that affine open piece $\mathcal{F}$ is flat over U. Therefore, let $X = \text{Spec}(B)$, let f make B into an A-algebra, and let $\mathcal{F}$ correspond to the B-module M. Then we shall prove:

(*) there is an element $f \in A$ such that $M_f = M \otimes_A A_f$ is a free A_f-module.

Note first that if

$$0 \to L \to M \to N \to 0$$

is an exact sequence of B-modules, and L_f is free over A_f, N_g is free over A_g, then M_{fg} is free over A_{fg}. To use this, recall that M being a B-module of finite type, admits a composition series:

$$(0) = M_0 \subset M_1 \subset M_2 \subset \dots \subset M_n = M$$

such that each factor M_{i+1}/M_i is isomorphic to $B/\wp_i$ for some prime ideal $\wp_i \subset B$ (BOURBAKI, Alg. Comm., Ch. 4, §1.4). Therefore it suffices to prove (*) for these $B/\wp_i$ and then it is proven for any M.

Therefore we may assume $M = B$, and B is an integral domain. Let K be the quotient field of A, and L the quotient field of B. We shall prove (*) by induction on the transcendence degree n of L over K. First, apply Noether's normalization lemma to the K-algebra $B \otimes_A K$; it follows that there exist n elements $f_1, \ldots, f_n \in B$ such that $B \otimes_A K$ is integral over the polynomial ring $K[f_1, \ldots, f_n]$. Then although B is not necessarily integral over $A[f_1, \ldots, f_n]$, there are only a finite number of denominators occurring in the relations of integral dependence of the generators of B over $K[f_1, \ldots, f_n]$: Therefore, for some $f \in A$,

$$(\#) \qquad B_f \text{ is integral over } A_f[f_1, \ldots, f_n] .$$

Then B_f is an $A_f[f_1, \ldots, f_n]$-module of finite type: consequently, we can find m elements $c_1, \ldots, c_m \in B_f$ generating a free $A_f[f_1, \ldots, f_n]$-submodule of B_f, such that the quotient is a torsion module

$$0 \to A_f[f_1, \ldots, f_n]^m \to B_f \to D \to 0 .$$

Now $A_f[f_1, \ldots, f_n]^m$ is clearly a free A_f-module, so it suffices to prove (*) for D. But, finally, replacing D by the quotients of a sufficiently fine composition series, we are reduced to proving (*) for integral A-algebras B' of transcendence degree <u>less than</u> <u>n</u> over A.

QED

3° We are left with the general case; a coherent $\mathcal{F}$ on $\mathbf{P}_n \times S$. Let p be the projection from $\mathbf{P}_n \times S$ to S, and put:

$$\mathcal{E}_m = p_*(\mathcal{F}(m)) .$$

As a first step, we note:

(*) there is a finite set of locally closed subsets $Y_1, \ldots, Y_k$ of S such that $S = \cup Y_i$, and such that if Y_i is given its reduced subscheme structure, $\mathcal{F} \otimes_{\underline{o}_S} \underline{o}_{Y_i}$ is flat over Y_i.

<u>Proof</u>: Immediate by 2° and the d.c.c. for closed subsets of S. From this we conclude several simplifying facts:

(i) there is a <u>uniform</u> m_0 such that if $m \geq m_0$, then for <u>all</u> $s \in S$, $H^i(\mathbf{P}_{n,s}, \mathcal{F}_s(m)) = (0)$, for $i > 0$ (notations as in Lecture 7) and $\mathcal{E}_m \otimes \mathcal{K}(s)$ is isomorphic to $H^0(\mathbf{P}_{n,s}, \mathcal{F}_s(m))$.

<u>Proof:</u> Put together (*); §7, 2° part (ii) applied over the base schemes Y_i; §7, 3°, Corollary $1\frac{1}{2}$; and 3°, part (i) applied to the inclusion $Y_i \subset S$.

(ii) Only a finite number of polynomials $P_1, \ldots, P_k$ occur as Hilbert polynomials of the sheaves $\mathcal{F}_s$ on the fibres $\mathbf{P}_{n,s}$ over S.

Fix m_0 as in (i), and let $g: T \to S$ be any base extension (T noetherian). Suppose first of all that $\mathcal{F}_g$ on $\mathbf{P}_n \times T$ is flat over T. Then by Corollary 2 in 3°, Lecture 7, the canonical map

$$g^*(\mathcal{E}_m) \to q_*(\underline{\mathcal{F}}_g(m)), \quad \text{for } m \geq m_0$$

is an isomorphism, and $g^*(\mathcal{E}_m)$ is locally free on T (where $q: \mathbf{P}_n \times T \to T$ is the projection). Conversely, suppose $g^*(\mathcal{E}_m)$ is flat, for all $m \geq m_0$: then by Corollary 3 in 3°, Lecture 7, $\mathcal{F}_g$ is flat over T.

Now any two stratifications of S have "g.c.d. stratification": i.e., given

$$S = \cup Y_i = \cup Z_j ,$$

then S is also the union of the locally closed subsets $W_{ij} = \text{Supp}(Y_i) \cap \text{Supp}(Z_j)$, and one can endow W_{ij} with a scheme structure by taking the sum of the sheaves of ideals defining Y_i and defining Z_j. By the result of 1°, each of the coherent sheaves $\mathcal{E}_m'$ has an associated flattening stratification. What we have just proven is that a flattening stratification for $\mathcal{F}$ is essentially the g.c.d. of the flattening stratifications of <u>all</u> $\mathcal{E}_m$ for $m \geq m_0$. To be precise, let $Y_e^{(m)}$ be the component of the flattening stratification of $\mathcal{E}_m$ on which $\mathcal{E}_m$ becomes locally free of rank e. Let $P_1, \ldots, P_k$ be the Hilbert polynomials of (ii). Then I claim that, for all i,

$$Z_i = \bigcap_{m=m_0}^{\infty} Y_{P_i(m)}^{(m)}$$

makes sense: Each finite intersection is, as just explained, a locally closed subscheme. But, set-theoretically,

$$\text{Supp } Z_i = \bigcap_{m=m_0}^{m_0+n} \text{Supp}\left(Y_{P_i(m)}^{(m)}\right) .$$

<u>Proof:</u> Let s be in $Y_{P_i(m)}^{(m)}$ for the n+1 values of m between m_0 and $m_0 + n$. Let P_j be the Hilbert polynomial of $\mathcal{F}_s$ on $\mathbf{P}_{n,s}$. Since the higher cohomology of $\mathcal{F}_s$ vanishes by (ii), we have

$$P_j(m) = \dim_{\mathcal{K}(s)} \mathcal{E}_m \otimes \mathcal{K}(s) = P_i(m) \quad .$$

But $P_i - P_j$ has degree at most n, and n+1 zeroes: therefore it is identically zero.

QED

Consequently, Z_i is the limit of a descending chain of locally closed subschemes with fixed support, i.e., of closed subschemes in a fixed open set U. By the d.c.c. for closed subschemes, it terminates and Z_i is actually a finite intersection which makes sense.

It is now trivial that $Z_1, \ldots, Z_k$ is a flattening stratification for $\mathcal{F}$ over S.

An obvious strengthening of the result is this:

<u>Corollary</u>: Let $X \xrightarrow{f} S$ be a morphism which can be factored:

$$\begin{array}{ccc} X & \overset{i}{\hookrightarrow} & \mathbf{P}_n \times S \\ & {\scriptstyle f}\searrow & \downarrow {\scriptstyle p_2} \\ & & S \end{array}$$

where i is a closed immersion. Let $\mathcal{F}$ be a coherent sheaf on X: then $\mathcal{F}$ defines a flattening stratification $\{Z_i\}$ on S.

Another important consequence of our method of proof is that the stratification $\{Z_i\}$ can be indexed by Hilbert polynomials P_i so that

i) the induced sheaf $\mathcal{F} \otimes_{\underline{o}_S} \underline{o}_{Z_i}$ has Hilbert polynomial P_i on $\mathbf{P}_n \times Z_i$,

ii) if $i \neq j$, then $P_i \neq P_j$.

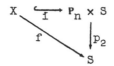

LECTURE 9

CARTIER DIVISORS

1° We assume that X is a noetherian scheme with structure sheaf $\underline{o}_X$.

<u>Definition-Proposition</u>: There is a unique sheaf $\underline{K}_X$ (of $\underline{o}_X$-modules) on X such that for affine open $U \subset X$,

$$\Gamma(U, \underline{K}_X) = \text{total quotient ring of } \Gamma(U, \underline{o}_X)$$

and for $U \subset V$, the restriction is the natural one.

<u>Proof</u>: Everything is easily reduced to this point: Say $U = \operatorname{Spec}(R)$, and $U_{f_i} = \operatorname{Spec} R_{(f_i)}$ are given where U_{f_i}, $1 \leq i \leq n$, form a covering of U: i.e., $1 \in (f_1, \ldots, f_n)$. Suppose $a_i, b_i \in R_{(f_i)}$, b_i not a 0-divisor in $R_{(f_i)}$, and assume $\{a_i/b_i \mid 1 \leq i \leq n\}$ agree on $U_i \cap U_j$, i.e., $b_j a_i - a_j b_j$ is 0 in $R_{(f_i f_j)}$. Then we must find α, $\beta \in R$, β not a 0-divisor in R such that $\alpha b_i - \beta a_i$ is 0 in $R_{(f_i)}$.

i) Multiplying a_i and b_i by f_i^N (for $N \gg 0$, and all i), we can assume that all elements a_i, b_i are in R, and that $a_i b_j = a_j b_i$ in R.

ii) Put $\mathfrak{A} = \{\beta \in R \mid \beta a_i \text{ is in the ideal } (b_i) \text{ in } R_{(f_i)}, \text{ all } i\}$.

Then one checks that $b_1, \ldots, b_n \in \mathfrak{A}$. Now say $c \in R$ and $c \cdot \mathfrak{A} = (0)$. Then $c \cdot b_i = 0$, all i. But b_i is a non-0-divisor in $R_{(f_i)}$, so c must go to 0 in $R_{(f_i)}$, i.e., $f_i^N \cdot c = 0$. Since $1 \in (f_1, \ldots, f_n)$, this implies that $c = 0$.

iii) But since R is noetherian, any $\mathfrak{A}$ with this property contains a non 0-divisor β. Now it follows that $\beta \cdot a_i/b_i$ is actually a section of $\underline{o}_X$ over U, hence for some $\alpha \in R$, $\beta \cdot a_i/b_i = \alpha$.

QED

We mention that $\underline{K}_X$ is not always quasi-coherent! Also, one checks that the stalks $\underline{K}_x$ of $\underline{K}_X$ are just the total quotient rings of the stalks $\underline{o}_x$. Finally, we can define $\underline{K}_X^*$ to be the subsheaf of <u>units</u> of the sheaf of rings $\underline{K}_X$, i.e.,

$$\Gamma(U, \underline{K}_X^*) = \text{invertible elements of } \Gamma(U, \underline{K}_X) \ .$$

Note that $\underline{o}_X \subset \underline{K}_X$ and $\underline{o}_X^* \subset \underline{K}_X^*$.

<u>Definition</u>: <u>A Cartier divisor</u> D on X is a section over X of $\underline{K}_X^*/\underline{o}_X^*$. More concretely, a Cartier divisor is given by a collection of elements

$$D_x \in \underline{K}_x^*/\underline{o}_x^*$$

such that, for all x, there is an open neighborhood U of X, and an element $f \in \Gamma(U, \underline{K}_X^*)$ which induces D_x for all $x \in U$. The element f will be called a <u>local equation</u> of D in U: It is unique up to a unit in $\underline{o}$. A Cartier divisor can be determined by specifying local equations $\{f_i\}$ with respect to an open covering $\{U_i\}$, so long as f_i/f_j is a unit in $U_i \cap U_j$.

Note that the set of all Cartier divisors froms a group. Although this law comes from <u>multiplying</u> local equations, we follow hallowed convention and write it additively: i.e., as $D_1 \pm D_2$ for the combination $f_1 \cdot f_2^{\pm 1}$ of local equations.

Associated to a Cartier divisor D is a coherent subsheaf:

$$\underline{o}_X(D) \subseteq \underline{K}_X$$

which is an invertible sheaf of $\underline{o}_X$-modules. Namely, for all x, put:

$$[\underline{o}_X(D)]_x = f_x^{-1} \cdot \underline{o}_x \subset \underline{K}_x$$

where f_x is the element of $\underline{K}_x$ induced by a local equation f of D. This is clearly independent of the choice of f, and, if f is a local equation in U, then

$$\underline{o}_X \mid U \xrightarrow[\text{mult. by } f^{-1}]{\sim} \underline{o}_X(D) \mid_U$$

is an isomorphism of sheaves of $\underline{o}_X$-modules.

It is not hard to check that this actually gives an isomorphism between the set of Cartier divisors on X, and the set of invertible coherent subsheaves of $\underline{K}_X$.

Definition: A Cartier divisor D is effective if equivalently:

i) its local equations f are sections of $\underline{o}_X$,

or ii) $\underline{o}_X \subset \underline{o}_X(D) \subset \underline{K}_X$,

or iii) $\underline{o}_X(-D)$ is a sheaf of ideals.

We shall write: $D > 0$ to mean D is effective. Suppose D is an effective Cartier divisor, and let $\underline{o}_D$ denote the cokernel:

$$(*) \qquad 0 \to \underline{o}_X(-D) \to \underline{o}_X \to \underline{o}_D \to 0$$

If one takes the structure sheaf $\underline{o}_D$ on the topological space which is the support of $\underline{o}_D$, one obtains a closed subscheme of X: By abuse of language, we shall also call this closed subscheme D. Since this closed subscheme determines its sheaf of ideals $\underline{o}_X(-D)$, which in turn determine local equations f in $\underline{o}_X$ (via $\underline{o}_X(-D) = f \cdot \underline{o}_X$), the Cartier divisor D is termined by the closed subscheme D and our confusion should not be dangerous.

Moreover, when $D > 0$, the image s of the section $1 \in \Gamma(X, \underline{o}_X)$ in $\Gamma(X, \underline{o}_X(D))$ will be called the global equation of D. In fact, if we let

$$\underline{o}_X(D) \xrightarrow[\sim]{\varphi} \underline{o}_X$$

be any isomorphism of modules, $\varphi(s)$ is a local equation for D at x. Moreover, in the exact sequence (*), the inclusion of $\underline{o}_X(-D)$ in $\underline{o}_X$ can be interpreted as tensoring with s.

A Cartier divisor D determines even more things:

Definition: The support of D is the closed subset consisting of those $x \in X$ at which 1 is not a local equation.

Definition: The divisor class associated to the Cartier divisor D is the element of Pic (X) obtained by the co-boundary:

$$H^0(X, \underline{K}^*/\underline{o}^*) \to H^1(X, \underline{o}^*) = \text{Pic}(X) \quad ,$$

via the exact sequence:

$$(\#)_C \qquad 0 \to \underline{o}_X^* \to \underline{K}_X^* \to \underline{K}_X^*/\underline{o}_X^* \to 0 \quad .$$

One checks immediately that this element of Pic (X) is, in fact, given by the invertible sheaf $\underline{o}_X(D)$.

Definition: Two Cartier divisors D_1, D_2 are linearly equivalent (written $D_1 \equiv D_2$) if, equivalently,

i) $\underline{o}_X(D_1) \cong \underline{o}_X(D_2)$, as $\underline{o}_X$-modules,

ii) the divisor class of D_1 equals the divisor class of D_2,

iii) there is an $f \in \Gamma(X, \underline{K}_X^*)$ such that

$$\begin{array}{ccc} f\cdot\underline{o}_X(D_1) & = & \underline{o}_X(D_2) \\ \cap & & \cap \\ & \underline{K}_X & \end{array}.$$

Definition: If $f \in \Gamma(X, \underline{K}_X^*)$, then the Cartier divisor with f as its local equation everywhere will be denoted (f). Such divisors are called principal, and by use of the exact sequence $(\#)_C$, one sees:

$$D_1 \equiv D_2 \text{ if and only if } D_1 = D_2 + (f), \quad \text{for some } f \in \Gamma(X, \underline{K}_X^*).$$

Next, suppose an invertible sheaf L is given—consider the set of all effective Cartier divisors D whose divisor class is L. That is to say, look for isomorphisms α:

$$\begin{array}{ccccc} & & L & & \\ & \varphi \nearrow & \downarrow \wr \alpha & & \\ \underline{o}_X & \subset & \underline{o}_X(D) & \subset & \underline{K}_X \end{array}.$$

Letting φ be the composition in the diagram, one sees conversely that for every injective homomorphism φ, there is a unique Cartier divisor D such that φ extends to an isomorphism α of $\underline{o}_X(D)$ and L. Thus D can be determined, for example, by letting $s = \varphi(1)$, and choosing local isomorphisms:

$$L\,|_{U_i} \xrightarrow{\sim} \underline{o}_X\,|_{U_i} .$$

Then the image of s in $\Gamma(U_i, \underline{o}_X)$ is a local equation for D. As above, we call $s \in \Gamma(X, L)$ a global equation for D. Note that the fact that φ is injective corresponds to the fact that s is not a 0-divisor. The above reasoning leads to:

Proposition: If L is an invertible sheaf, then there is a natural isomorphism:

$$\left\{\begin{array}{l}\text{effective Cartier divisors} \\ D \text{ s.t. } \underline{o}_X(D) \cong L\end{array}\right\} \xrightarrow{\sim} \left\{\begin{array}{l}\text{sections } s \in \Gamma(X, L)\text{, not} \\ \text{0-divisors, modulo} \\ s \sim \alpha\cdot s, \text{ for } \alpha \in \Gamma(X, \underline{o}_X^*)\end{array}\right\}$$

Example: Let $X = \operatorname{Proj} k[X_0, \ldots, X_n]$, k a field. Then as in Lecture 5, X carries the sheaf $\underline{o}_X(1)$, and there are homomorphisms:

$$\left\{\begin{array}{l}\text{vector space of homogeneous} \\ \text{forms in } X_0, \ldots, X_n \text{ of degree } d\end{array}\right\} \to \Gamma(X, \underline{o}_X(d)) .$$

Therefore, each form $F(X_0, \ldots, X_n)$ of degree d is the global equation of an effective Cartier divisor $D \subset X$ such that $\underline{o}_X(D) \cong \underline{o}_X(d)$. This d is called the hypersurface with equation F, (or, if $d = 1$, the hyperplane).

2° Cartier divisors are closely related to the concept of depth. If $z \in X$ is a point where depth $(\underline{o}_z) = 0$, then $\underline{K}_z = \underline{o}_z$, hence $(\underline{K}^*/\underline{o}^*)_z = (1)$, and every Cartier divisor is trivial in a neighborhood of z. The remarkable thing is that Cartier divisors are determined by their equations at points of depth 1 :

Proposition: Let X be a noetherian scheme, D_1, D_2 two C-divisors on X. Then $D_1 = D_2$ if and only if their images in the stalks $(\underline{K}^*/\underline{o}^*)_x$ are equal for all x where depth $(\underline{o}_x) = 1$.

Proof: It suffices to prove that the images $(D_1)_x$ and $(D_2)_x$ of D_1 and D_2 are equal in all stalks $(\underline{K}^*/\underline{o}^*)_x$. But, multiplying both by a suitable non-0-divisor in $\underline{o}_x$, this reduces to proving:

(*) Given two principal ideals I_1, I_2, generated by non-0-divisors, in a local noetherian ring $\mathcal{O}$, then $I_1 = I_2$ if $I_1(\mathcal{O})_{\wp} = I_2(\mathcal{O})_{\wp}$ for all localizations $(\mathcal{O})_{\wp}$ of depth 1.

But certainly $I_1 = I_2$ if $I_1(\mathcal{O})_{\wp} = I_2(\mathcal{O})_{\wp}$ for all prime ideals $\wp$ associated to I_1 or I_2. And if $\wp$ is associated to $I_1 = (a_1)$, then in $(\mathcal{O})_{\wp}$, a_1 is a non-0-divisor such that all non-units in $(\mathcal{O})_{\wp}/a_1\cdot(\mathcal{O})_{\wp}$ are 0-divisors: i.e., depth $(\mathcal{O}_{\wp}) = 1$.

QED

In a very similar way, it can be proved that a Cartier divisor D is effective if and only if it is effective at all points x, where depth $(\underline{o}_x) = 1$.

Corollary: Let X be a normal noetherian scheme, i.e., all local rings $\underline{o}_x$ are integrally closed domains. Then two Cartier-divisors D_1, D_2 are equal if and only if they are equal at all points x of codimension 1.

Proof: By the principal ideal theorem, a normal local ring of Krull dimension ≥ 2 has depth ≥ 2.

QED

Now assume for the rest of 2° that X is an irreducible normal noetherian scheme. If K is the stalk of $\underline{o}_X$ at the generic point of X, then $\underline{K}_X$ is simply the constant sheaf:

$$\Gamma(U, \underline{K}_X) = K, \quad \text{all } U .$$

Incidentally, this proves immediately that $H^1(X, \underline{K}_X^*) = (0)$, hence by the exact sequence $(\#)_C$ (1°): every invertible sheaf $\mathfrak{L}$ on X is the divisor class of some Cartier-divisor.

Definition: A Weil divisor on X is a formal sum

$$\sum_{i=1}^{n} r_i E_i$$

where $E_1, \ldots, E_n$ are closed irreducible subsets of codimension 1.

If, for all $x \in X$ of codimension 1, we define a sheaf Z_x by:

$$\Gamma(U, Z_x) = \begin{cases} (0) & \text{if } x \notin U \\ Z & \text{if } x \in U \end{cases}$$

then one checks that a Weil divisor is the same thing as a section of the sheaf

$$\bigoplus_{x \text{ of codim } 1} Z_x .$$

Now there is a canonical exact sequence:

$$(\#)_W \qquad 0 \to \underline{o}_X^* \to \underline{K}_X^* \longrightarrow \bigoplus_{x \text{ of codim } 1} Z_x .$$

Namely, given $f \in \Gamma(U, \underline{K}_X^*) = K^*$, define its image to be:

$$\sum_{\substack{x \in U \\ x \text{ of codim } 1}} \operatorname{ord}_x(f) \cdot \{\overline{x}\} ,$$

where $\operatorname{ord}_x(f)$ is the order of f at x. In other words, say $U = \operatorname{Spec}(R)$. Then let $f = g/h$, where $g, h \in R$, and let

$$(g) = \wp_1^{(s_1)} \cap \wp_2^{(s_2)} \cap \ldots \cap \wp_n^{(s_n)}, \quad s_i \geq 0$$

$$(h) = \wp_1^{(t_1)} \cap \wp_2^{(t_2)} \cap \ldots \cap \wp_n^{(t_n)}$$

where the $\wp_i$ are minimal prime ideals, and $\wp^{(t)}$ is the tth "symbolic" power of $\wp$ $[\wp^{(t)} = R \cap (\wp \cdot R_\wp)^t]$. Then the image of f is:

$$\sum_{i=1}^{n} (s_i - t_i) \{\text{closure of point given by } \wp_i\} .$$

Note that if $s_i = t_i$ for all i, then $(g) = (h)$, hence f is a unit in R: this shows that $(\#)_W$ is exact.

Putting $(\#)_C$ and $(\#)_W$ together, we obtain an inclusion

$$\underline{K}^*/\underline{o}^* \subset \oplus_x Z_x ,$$

hence the group of Cartier divisors is embedded in the group of Weil divisors. This is, in fact, just an interpretation of the Corollary just above: for if $x \in X$ has codimension 1, and (π) is the maximal ideal in $\underline{o}_x$, then the stalk of a Cartier divisor at x has a local equation of the form π^r, for a well determined integer r. The corresponding Weil divisor is then just the sum over x of $r \cdot \{\overline{x}\}$.

Proposition: The group of Cartier divisors equals the group of Weil divisors if and only if all local rings $\underline{o}_x$ are UFD's; e.g., if X is a regular scheme.

Proof: The two types of divisors are equal if and only if the homomorphism of stalks in $(\#)_W$:

$$(\underline{K}_X^*)_y \to \left[\bigoplus_{x \text{ of codim } 1} Z_x\right]_y$$

is surjective. But this is simply:

$$K^* \to \bigoplus_{\substack{\wp \subset \underline{o}_X \\ \text{minimal primes}}} Z$$

assigning to $f = g/h$ the difference of the orders of g and h at all $\wp$. This is surjective if and only if every $\wp \subset \underline{o}_X$ is a principal ideal: i.e., if and only if $\underline{o}_X$ is a UFD.

LECTURE 10

FUNCTORIAL PROPERTIES OF EFFECTIVE CARTIER DIVISORS

1° The simplest operation to perform with Cartier divisors is to take inverse images: say $X \xrightarrow{g} Y$ is a morphism of noetherian schemes, and say D is an effective C-divisor on Y. Then it is quite clear what $g^*(D)$ ought to mean: Fix an open covering $\{U_i\}$ of Y and local equations f_i for D in U_i, where $f_i \in \Gamma(U_i, \underline{o}_Y)$. Then $g^*(D)$ should be defined by local equations $g^*(f_i)$ in the open covering $g^{-1}(U_i)$. However, $g^*(f_i)$ can be a 0-divisor, even 0. The best thing is to assume:

$$(*) \qquad \text{for all } x \in A(X), \quad g(x) \notin \text{Supp}(D) .$$

Then $g^*(f_i)$ is not a 0-divisor, and $g^*(D)$ makes sense.

Proof: Suppose $a \cdot g^*(f_i) = 0$, where $a \in \underline{o}_x$, and $x \in X$. Then let x' be the generic point of some component of the support of the section a of $\underline{o}_X$ (defined near x): We may take

$$x' \in \text{Spec}(\underline{o}_x) \subset X .$$

Then $\underline{o}_{x'}$ has depth 0 since the induced element $a' \in \underline{o}_{x'}$ is killed by a power of the maximal ideal $m_{x'}$ (cf. Lecture 8, 2°), and since $a' \neq 0$. But then $x' \in A(X)$, hence $g(x') \notin \text{Supp}(D)$. Therefore, the local equation f_i for D is a unit at $g(x')$; therefore $g^*(f_i)$ is a unit at x'. Therefore, in $\underline{o}_{x'}$:

$$a' = [a' \cdot g^*(f_i)] \cdot g^*(f_i)^{-1} = 0 .$$

This contradiction proves the result.

Note that if g is flat, (*) is automatic. For if g is flat, then for all $x \in A(X)$, $g(x) \in A(Y)$, (Lecture 6), hence $g(x)$ is not in the support of any C-divisor (Lecture 8, 2°).

2° A more interesting question is when can one define a direct image $g_*(D)$ of an effective C-divisor D on X. In this section, we treat the "elementary" case:

g is finite and flat.

Then g_* can be defined by Norms! The problem is essentially algebraic, since it is local on Y: let $U = \text{Spec}(A)$ be an open affine subset of Y, and let $g^{-1}(U) = \text{Spec}(B)$. Then B is an A-algebra, which is of finite type as A-module. Moreover, since $g_*(\underline{o}_X)$ is a locally free sheaf on Y, if we take U sufficiently small, B is a free A-module too. We are then set up for norms:

if $\beta \in B$, let $T_\beta: B \to B$ be multiplication by β.
if $b_1, \ldots, b_n$ are a basis of B over A, let

$$T_\beta(b_i) = \sum_{j=1}^{n} a_{ij} b_j \ .$$

Then:

$$\text{Nm}(\beta) = \det(a_{ij}) \ .$$

This is naturally independent of the basis b_i, and has the obvious properties:

$$\text{Nm}(\beta_1 \cdot \beta_2) = \text{Nm}(\beta_1) \cdot \text{Nm}(\beta_2)$$
$$\text{Nm}(\alpha) = \alpha^n, \text{ if } \alpha \in A \ .$$

Although the norm is not always a product of β and its conjugates, at least one has:

(*) for all β, there is a β' such that $\text{Nm}(\beta) = \beta \cdot \beta'$.

<u>Proof</u>: Let $P(X) = \det(X \cdot \text{identity} - T_\beta)$ be the characteristic polynomial of T_β. Then (Cayley-Hamilton theorem) $P(T_\beta) = 0$, hence $P(\beta) = P(T_\beta)(1) = 0$, or, writing out P:

$$\beta^n + a_1\beta^{n-1} + \ldots + a_{n-1} \cdot \beta + \text{Nm}(\beta) = 0.$$

QED

One also has the important:

(**) If $\beta \in B$ is not a 0-divisor, then $\text{Nm}(\beta)$ is not a 0-divisor.

<u>Proof</u>: We use a simple general fact:

<u>Lemma A</u>: Let $X \xrightarrow{g} Y$ be a finite flat morphism of noetherian schemes. Let $x \in X$. If $g(x)$ has depth 0, then x has depth 0, and conversely.

<u>Proof</u>: If depth $g(x) = 0$, then there exists $a \in \underline{o}_{g(x)}$, $a \neq 0$, whose annihilater is $m_{g(x)}$, the maximal ideal. Since g is flat, $g^*: \underline{o}_{g(x)} \to \underline{o}_x$ is injective and $g^*(a) \in \underline{o}_x$ is not 0. Since g is finite, $m_{g(x)} \cdot \underline{o}_x$ is primary for the maximal ideal m_x: since $m_{g(x)} \cdot \underline{o}_x$ kills $g^*(a)$, the depth of x is 0. The converse was proven in Lecture 6.

QED

Returning to B/A: Suppose $Nm(\beta)$ is a 0-divisor. Then there is a prime ideal $\mathfrak{p} \subset A$ such that depth $(A_{\mathfrak{p}}) = 0$, and such that $Nm(\beta)$ is a 0-divisor in $A_{\mathfrak{p}}$ [i.e., let $a \cdot Nm(\beta) = 0$, and let $\mathfrak{p}$ be a minimal prime ideal containing the annihilater of a]. Replace A by $A_{\mathfrak{p}}$ and B by $B_{\mathfrak{p}} = B \otimes_A A$. Then B is a semi-local ring all of whose localizations have depth 0 by the lemma. Then if β is not a 0-divisor, β is in none of the maximal ideals of B, i.e., β is a unit in B. Since Nm is multiplicative, $Nm(\beta)$ is a unit too, which contradicts our assumption.

To apply the norm to the definition of g_*, we need:

Lemma B: Let $X \xrightarrow{g} Y$ be a finite morphism of noetherian schemes, and let L be an invertible sheaf on X. Then there exists an open covering $\{U_i\}$ of Y such that L is isomorphic to $\underline{o}_X$ in each open set $g^{-1}(U_i)$.

Proof: For all $y \in Y$, look at the module $M = g_*(L)_y$ over $B = g_*(\underline{o}_X)_y$. Since g is finite, B is a semi-local ring, and if $\overline{\mathfrak{A}}$ is its radical,

$$B/\overline{\mathfrak{A}} \cong \bigoplus_{x \text{ over } y} \overline{\kappa}(x) .$$

Therefore, $M/\mathfrak{A} \cdot M$ is certainly free of rank 1: hence M is free of rank 1 over B (cf. BOURBAKI, Alg. Comm., Ch. II, §3, Prop. 5). Let μ_y be a basis of M; then, μ_y is induced by a section μ of $g_*(L)$ in an open neighborhood U_1 of y. Multiplication by μ defines a homomorphism:

$$g_*(\underline{o}_X) \xrightarrow{\mu} g_*(L)$$

in U_1. The kernel and cokernel are coherent sheaves on Y whose stalks at y are (0): therefore, both are (0) in a whole neighborhood $U_2 \subset U_1$ of y. Then in $g^{-1}(U_2)$, multiplication by μ gives an isomorphism of $\underline{o}_X$ and L.

QED

Now in our case, we are given an effective C-divisor D on X: By the lemma, there is an open affine covering $U_i = \text{Spec}(A_i)$ of Y such that D is principal in $g^{-1}(U_i) = \text{Spec}(B_i)$. Therefore D is defined by an equation $\beta_i \in B_i$, for all i, β_i not a 0-divisor. One checks that $\beta_i \cdot \beta_j^{-1}$ is a unit in $\Gamma(g^{-1}(U_i \cap U_j), \underline{o}_X)$, hence $Nm(\beta_i) \cdot Nm(\beta_j)^{-1}$ is a unit in $\Gamma(U_i \cap U_j, \underline{o}_Y)$. Therefore, the sections $Nm(\beta_i)$ define a C-divisor $g_*(D)$.

3° Remarkably, the direct image $g_*(D)$ can be defined in a very much more general case: 2° is really just "case 0" in an infinite set of cases, in each of which $g_*(D)$ can be defined, but requiring, in each successive case, the computation of one more determinant, among other things. We have in mind the following situation:

$$\begin{array}{ccccc} P_n \times Y & \supset & X & \supset & V \\ & \searrow^{p_1} & \downarrow g & & \downarrow g_0 \\ & & Y & \supset & U \end{array}$$

where (a) X is a closed subscheme of $\mathbf{P}_n \times Y$, U is open in Y,

(b) $V = g^{-1}(U)$, g_0 is the restriction of g,

(c) g_0 is finite,

(d) g is of finite Tor-dimension,

(e) all points $y \in Y$, where $\underline{o}_y$ has depth 0 or 1, are in U.

Then in this situation there is a natural definition of $g_*(D)$. (Cf. Mumford, Geometric Invariant Theory, Ch. 5, §3.) In fact, if g_0 is also flat, $g_*(D)$ is uniquely determined by the requirement:

$$g_*(D)|_U = g_{0,*}(D|_V).$$

4° In this section, I want to define the concept of a relative (effective) C-divisor. Suppose $X \xrightarrow{f} Y$ is a flat morphism of finite type of noetherian schemes. The question is, when should a divisor $D \subset X$ be regarded as a family of C-divisors on the various fibres of f.

Proposition-Definition: An effective C-divisor $D \subset X$ is said to be a relative divisor over Y if equivalently:

i) D is flat $/Y$,

or ii) for all $x \in X$, the local equation F of D at x is not a zero-divisor in the ring $\underline{o}_x \otimes_{\underline{o}_y} \mathcal{K}(y)$, where $y = f(x)$,

or iii) for all $y \in Y$,

$$A(f^{-1}(y)) \cap \mathrm{Supp}\,(D) = \emptyset .$$

Proof: (ii) and (iii) are obviously equivalent. To prove them equivalent to (i), pass to the algebraic setup, since the problem is local on X and Y: then one has B, a flat A-algebra, and $F \in B$ a non-0-divisor. Let $\mathfrak{p} \subset A$ be a prime ideal. Since B is flat $/A$, $B/\mathfrak{p} \cdot B$ is flat over $A/\mathfrak{p}$: therefore all prime ideals $\overline{\mathfrak{q}} \subset B/\mathfrak{p} \cdot B$ associated to (0) contract to prime ideals in $A/\mathfrak{p}$ associated to (0), i.e., contract (0) itself since $A/\mathfrak{p}$ is an integral domain (this is Example 1, Lecture 6). In other words, all prime ideals $\mathfrak{q} \subset B$ associated to $\mathfrak{p} \cdot B$ satisfy $\mathfrak{q} \cap A = \mathfrak{p}$. Therefore, all such $\mathfrak{q}$ are associated to (0) in $B \otimes$ [quotient field of $A/\mathfrak{p}$], i.e., such $\mathfrak{q}$ correspond to $x \in A(f^{-1}(y))$ if y corresponds to $\mathfrak{p}$. Therefore, hypothesis (iii) asserts:

iii)* F is not in any associated prime ideal of $\mathfrak{p} \cdot B$, for any prime ideal $\mathfrak{p} \subset A$.

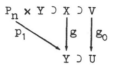

To prove this is equivalent to the flatness of $B/F \cdot B$ over A, recall that flatness is equivalent to:

$$\mathrm{Tor}_1^A (B/F \cdot B,\ A/\wp) = (0)\ ,$$

all prime ideals $\wp \subset A$, (this is easy— cf. BOURBAKI, Comm. Alg., Ch. I, §4). But using:

$$\mathrm{Tor}_1^A (B,\ A/\wp) \longrightarrow \mathrm{Tor}_1^A (B/F \cdot B,\ A/\wp) \to B/\wp \cdot B \xrightarrow{F} B/\wp B$$

and the flatness of B over A, the vanishing of this Tor is equivalent to (iii)*.

QED

The important point concerning relative Cartier divisors is this: given a fibre product situation:

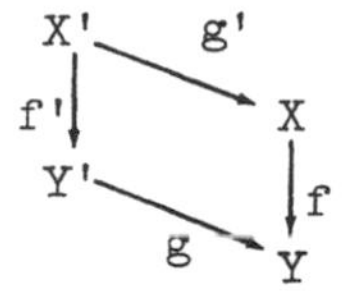

and an effective C-divisor D in X, relative to f, then $g'^*(D)$ is always defined. For, by the remarks at the end of Lecture 6, a point $x' \in A(X')$ is also in $A(f'^{-1}(y'))$, if $y' = f'(x)$. And

$$f'^{-1}(y') \cong f^{-1}(y) \underset{\mathrm{Spec}\ \mathcal{H}(y)}{\times} \mathrm{Spec}\ \mathcal{H}(y')$$

where $y = g(y')$. Therefore $f'^{-1}(y')$ is flat over $f^{-1}(y)$, hence $g'(x') \in A(f^{-1}(y))$. Therefore

$$g'(x') \notin \mathrm{Supp}\ (D)\ .$$

This implies that $g'^*(D)$ is defined. (Cf. 1°).

In particular, one can take $Y' = \mathrm{Spec}\ \mathcal{H}(y)$ for various $y \in Y$, and one obtains a family of C-divisors on the fibres $f^{-1}(y)$ of f —as required!

LECTURE 11

BACK TO THE CLASSICAL CASE

After spending so long in the arid generality of arbitrary noetherian schemes we return to our proper program—to investigate the set of curves on a given surface. In this lecture, we simply set the stage for working over a field k, recalling without proof some of the basic facts:

> Fix, once and for all, an algebraically closed field k. Recall, an algebraic scheme /k is a scheme X of finite type over k. All schemes, henceforth, will be algebraic schemes, and all functors will be functors on the category of algebraic schemes. Recall, a variety /k is a reduced and irreducible scheme /k. From now on, $\mathbf{P}_n$ will denote Proj $k[X_0, \ldots, X_n]$, (not Proj $\mathbf{Z}[X_0, \ldots, X_n]$).

(I.) Recall also the main result of dimension theory in this case (cf. ZARISKI-SAMUEL, vol. 2, p. 193):

(*) If X is an irreducible scheme, there is an integer n, the dimension of X, such that

$$\text{Krull dim } (\underline{o}_x) + \text{trans. deg. } \mathcal{H}(x)/k = n$$

for all $x \in X$.

Definition: If X is any scheme, let dim (X) be the maximum of the dimensions of the components of X.

It can be shown that dim (X) is also the cohomological dimension of X: thus if $i > \dim X$, $H^i(X, \mathcal{F}) = (0)$ for all sheaves $\mathcal{F}$ (cf. GODEMENT, Theorie des faisceaux, p. 197).

(II.) Definition: A scheme X is projective (resp. quasi-projective) if it is isomorphic to a closed (resp. locally closed) subscheme of $\mathbf{P}_n$ (for some n).

Definition: An invertible sheaf L on a scheme X is very ample if there exists an immersion

$$\varphi: X \to \mathbf{P}_n$$

(for some n) such that $\varphi^*(\underline{o}(1)) \cong L$.

There are several important remarks to make about this concept:

a) Suppose more generally that $L \cong \varphi^*(\underline{o}(1))$ for any morphism $\varphi: X \to \mathbf{P}_n$ at all. Then the induced sections $s_i = \varphi^*(X_i)$ of L span L. Conversely, if L is spanned by its global sections, one can choose a finite set $s_0, s_1, \ldots, s_n$ of sections which span L. Then $(L; s_0, \ldots, s_n)$ defines an X-valued point of $\mathbf{P}_n$, i.e., a morphism $\varphi: X \to \mathbf{P}_n$, such that $\varphi^*(\underline{o}(1)) = L$. In particular, a very ample sheaf is spanned by its global sections.

b) Suppose $H^0(X, L)$ is finite-dimensional, e.g., suppose X is a projective scheme. Then if L is spanned by its sections, there is a nearly canonical morphism $\varphi: X \to \mathbf{P}_n$ such that $L \cong \varphi^*(\underline{o}(1))$: namely, take a basis $s_0, s_1, \ldots, s_n$ of $H^0(X, L)$. These cannot all vanish at any one point, so $(L; s_0, \ldots, s_n)$ defines such a φ. More functorially, this defines a morphism:

$$\varphi: \quad X \to \mathbf{P}[H^0(X, L)] .$$

Note that in this embedding, $\varphi(X)$ is not contained in any hyperplane (in the scheme-theoretic sense: i.e., φ does not factor through a hyperplane $H \subset \mathbf{P}_n$). For if this happened, then for suitable $\alpha_0, \alpha_1, \ldots, \alpha_n$, one would have

$$\varphi^*(\Sigma\, \alpha_i X_i) \quad = \quad \Sigma\, \alpha_i s_i \quad = \quad 0$$

in $H^0(X, L)$, contradicting the independence of the s_i.

Definition: An invertible sheaf L on a scheme X is ample if there exists a positive integer n such that L^n is very ample.

(III.) An important fact about projective varieties X is that:

$$\Gamma(X, \underline{o}_X) \cong k .$$

This follows because, in any case, the ring $A = \Gamma(X, \underline{o}_X)$ is a finite dimensional commutative algebra over k. And since k is algebraically closed, if $k \subsetneq A$, then A contains 0-divisors. But since X is reduced and irreducible, even $\underline{o}_X$ contains no 0-divisors.

If X is any projective scheme, the finite-dimensional vector spaces $H^i(X, \underline{o}_X)$ are important invariants of X. One of the most interesting is the alternating sum of their dimensions, $\chi(\underline{o}_X)$. For historical reasons, when X is a projective variety of dimension n, one drops the term $\dim H^0(X, \underline{o}_X) = 1$, and counts down from H^n, obtaining the so-called arithmetic genus:

$$\begin{aligned} p_a(X) &= \dim H^n(X, \underline{o}_X) - \dim H^{n-1}(X, \underline{o}_X) + \ldots + (-1)^{n-1} \dim H^1(X, \underline{o}_X) \\ &= (-1)^n \{\chi(\underline{o}_X) - 1\} . \end{aligned}$$

This has the advantage that when X is a curve,

$$p_a(X) = \dim H^1(X, \underline{o}_X) = \text{usual genus of } X .$$

On the other hand, when X is a surface we get

$$p_a(X) = \dim H^2(X, \underline{o}_X) - \dim H^1(X, \underline{o}_X) .$$

[The point is that the Italians regarded $\dim H^2(X, \underline{o}_X)$ as the dominant term, and called it, for non-singular surfaces, the geometric genus, $p_g(X)$; while $\dim H^1(X, \underline{o}_X)$ was considered a "correction" term, and was called the irregularity $q(X)$. The reason is that for surfaces in $\mathbf{P}_3$ — which were looked at first— $q = 0$ and $p_a = p_g$.] In any case, for any projective scheme X we shall make the definition:

$$p_a(X) = (-1)^{\dim X} \{\chi(\underline{o}_X) - 1\} .$$

(IV.) A theorem which one can only use without thinking twice when the base is a field is the Künneth formula. This simple but convenient tool has grown to really awe-inspiring size in Grothendieck's tome (cf. EGA, §6, esp. Th. 6.7.3), but for our modest needs the following suffices:

For any schemes X, Y, let $\mathcal{F}, \mathcal{G}$ be quasi-coherent sheaves on X, Y respectively, then:

$$H^n(X \times Y, p_1^*\mathcal{F} \otimes p_2^*\mathcal{G}) \cong \bigoplus_{i+j=n} [H^i(X, \mathcal{F}) \otimes H^j(Y, \mathcal{G})],$$

(Proof by Czech-cohomology, and theorem of Eilenberg-Zilber.)

A Corollary of this is:

$$p_{1,*}[p_1^*\mathcal{F} \otimes p_2^*\mathcal{G}] \cong \mathcal{F} \otimes_k H^0(Y, \mathcal{G})$$

i.e., apply the Künneth formula to $U \times Y$, for $U \subset X$ affine and open. In particular, in view of (I.):

$$p_{1,*}[\underline{o}_{X\times Y}] \cong \underline{o}_X$$

if Y is a variety.

(V.) Definition: A variety X is non-singular if all local rings $\underline{o}_x$ are regular, $x \in S$.

Definition: A variety X is normal if all the local rings $\underline{o}_x$ are integrally closed, $x \in X$.

The product of non-singular varieties is non-singular; more generally, if $X \xrightarrow{f} Y$ is a flat surjective morphism with non-singular fibres, then X is non-singular if and only if Y is non-singular. [A flat morphism with non-singular fibres is known as a simple or "lisse" or smooth morphism.]

Moreover, the product of two reduced schemes is reduced; more generally, if $X \xrightarrow{f} Y$ is a flat surjective morphism with reduced fibres, then X is reduced if and only if Y is reduced. A simple consequence of the former is that for any algebraic schemes X and Y:

$$(X \times Y)_{red} \cong X_{red} \times Y_{red} \quad .$$

If X is an algebraic scheme, the set of all $x \in X$ such that $\underline{o}_x$ is regular is an open subset $U \subset X$. In particular, if X is a variety, and $x \in X$ is its generic point, then $\underline{o}_x$ is a field, hence regular; therefore there is an open dense subset $U \subset X$ which is non-singular.

(VI.) Finally we want to recall the Riemann-Roch theorem for curves which is the fundamental result describing the geometry on a curve.

<u>Definition.</u> A <u>curve</u> X is a 1-dimensional projective scheme all of whose closed points have depth 1, i.e., all its local rings are Cohen-Macauley. If $D \subset X$ is an effective Cartier divisor on X, let f_x be a local equation of D at $x \in X$. For all but a finite set of points, say $x_1, \ldots, x_n$, we may assume that $f_x = 1$. Then one defines

a) $$\deg (D) = \sum_{i=1}^{n} \dim_k[\underline{o}_{x_i}/(f_{x_i})] \quad .$$

Note that $\underline{o}_D$ is just $\underline{o}_{x_i}/(f_{x_i})$ at x_i, and (0) elsewhere, so that:

b) $$\deg (D) = \dim H^0(X, \underline{o}_D) \quad .$$

If X is non-singular, and (t_i) is the maximal ideal at x_i, then let

$$f_{x_i} = (\text{unit}) \cdot x_i^{r_i} \quad .$$

Then D is the Weil divisor $\Sigma_{i=1}^n r_i \cdot x_i$, and

c) $$\deg (D) = \sum_{i=1}^{n} r_i \quad .$$

The interesting thing about this invariant is that it depends only on the divisor class of D, not on D itself. This can be seen by using the definition (b) and the exact sequence

$$0 \to \underline{o}_X(-D) \to \underline{o}_X \to \underline{o}_D \to 0 \quad .$$

d) $$\deg (D) = \chi(\underline{o}_X) - \chi(\underline{o}_X(-D)) \quad .$$

Using this formula, one can extend the definition to arbitrary invertible sheaves L

e) $\deg(L) = \chi(\underline{o}_X) - \chi(L^{-1})$.

Riemann's half of the Riemann-Roch theorem then asserts simply:

THEOREM 1: a) $\deg(L \otimes M^{\pm 1}) = \deg L \pm \deg M$ hence

b) $\dim H^0(L) - \dim H^1(L) = \deg(L) + \chi(\underline{o}_X)$,

$= \deg(L) + 1 - p_a(X)$.

In other words, degree gives a homomorphism:

$$\mathrm{Pic}(X) \xrightarrow{\deg} \mathbf{Z}.$$

[If X is irreducible, then the kernel will be called $\mathrm{Pic}^\tau(X)$, and it is well-known to be canonically isomorphic to the group of k-rational points on a group-scheme—the so-called Jacobian variety of X. We shall have much more to say about this below.]

Roch's half of the Riemann Roch theorem tells how to compute the H^1 in terms of H^0.

THEOREM 2: There is a canonical coherent sheaf ω_X on X such that the vector spaces

$$H^1(X, L) \text{ and } H^0(X, \omega_X \otimes L^{-1})$$

and the vector spaces

$$H^0(X, L) \text{ and } H^1(X, \omega_X \otimes L^{-1})$$

are canonically dual to each other (for any invertible sheaf L). In particular, $\chi(L) = -\chi(\omega_X \otimes L^{-1})$.

[For a proof when X is reduced and irreducible, cf. SERRE, Groupes algébriques et ..., Ch. 4; in the general case, cf. Grothendieck's talk at the Bourbaki Seminar, exposé 149 and Hartshorne's forthcoming notes on Duality in the Springer Lecture Note Series. Actually, the proof here is quite simple. One Chooses an embedding

$$X \subset \mathbf{P}_n, \quad (\text{some } n) .$$

Then put $\omega_X = \underline{\mathrm{Ext}}^{n-1}_{\underline{o}_{\mathbf{P}_n}}[\underline{o}_X, \underline{o}_{\mathbf{P}_n}(-n-1)]$. Then one uses the standard results on change of rings in Ext's, the connections between H^i and $\underline{\mathrm{Ext}}^i$ in the general theory of sheaves—cf. Grothendieck's Tohoku paper, or Godement's book §7.3—and finally the last theorem in Serre's paper FAC: viz., if $\mathcal{F}$ is any coherent sheaf on $\mathbf{P}_n$, then

$$H^i(\mathbf{P}_n, \mathcal{F}) \text{ and } \mathrm{Ext}^{n-i}_{\underline{o}_{\mathbf{P}_n}}(\mathcal{F}, \underline{o}_{\mathbf{P}_n}(-n-1))$$

are canonically dual.

One further point which we will need: If X is reduced and irreducible, then ω_X is torsion-free and of rank 1 as $\underline{o}_X$-module. This can be seen in Serre's book, or by computing the $\underline{\mathrm{Ext}}^{n-1}$ above.]

The following consequence is the prototype of a large class of useful results: the <u>vanishing</u> <u>theorems</u>:

<u>Corollary</u>: Let L be an invertible sheaf on a reduced and irreducible curve X. Assume

$$\deg(L) > 2p_a(X) - 2 .$$

Then $H^1(X, L) = (0)$.

<u>Proof</u>: Suppose $H^1(X, L) \neq (0)$. Then

$$\begin{aligned} \dim H^1(X, \omega_X \otimes L^{-1}) &= \dim H^0(X, L) \\ &= \deg(L) + 1 - p_a(X) + \dim H^1(X, L) \\ &\geq p_a(X) + 1 \end{aligned}$$

while

$$\dim H^0(X, \omega_X \otimes L^{-1}) = \dim H^1(X, L) \geq 1 .$$

Let σ be a section of $\omega_X \otimes L^{-1}$; σ defines a homomorphism h:

$$\underline{o}_X \xrightarrow{h} \omega_X \otimes L^{-1} .$$

If h is not injective, then h annihilates a whole coherent sheaf of ideals. Since X is reduced and irreducible, a non-zero coherent sheaf of ideals is isomorphic to $\underline{o}_X$ at all but a finite set of points. Therefore, if h is not injective, the support of σ is 0-dimensional and ω_X would have torsion. Therefore, we get an exact sequence:

$$0 \to \underline{o}_X \xrightarrow{h} \omega_X \otimes L^{-1} \longrightarrow \kappa \to 0 .$$

Moreover, since the rank of $\omega_X \otimes L^{-1}$ is 1, the homomorphism h':

$$0 \to \underline{K}_X \xrightarrow{h'} (\omega_X \otimes L^{-1}) \otimes \underline{K}_X \to \kappa \otimes \underline{K}_X \to 0$$

is an isomorphism; hence $\kappa \otimes \underline{K}_X = (0)$, and κ is a torsion sheaf. Therefore we get the exact sequence:

$$H^1(X, \underline{o}_X) \to H^1(X, \omega_X \otimes L^{-1}) \to H^1(X, \kappa) = (0) .$$

Since $\dim H^1(X, \underline{o}_X) = p_a$, this is a contradiction.

QED

A further development of the theory shows that there is another constant n_0, depending only on X, such that when the degree of the invertible sheaf L is at least n_0, then L is very ample (cf. MATSUSAKA-MUMFORD, <u>Am</u>. <u>J</u>. <u>Math</u>., 1964). This gives the elegant Corollary: L is ample if and only if $\deg(L) > 0$. [To show that "$\deg L > 0$" is necessary, use Serre's theorem that "L ample" implies $H^1(L^n) = (0)$, for large n, hence $\chi(L^n) \to \infty$ as $n \to +\infty$, hence, by Theorem 1 $\deg(L) > 0$.]

Finally, there is a third part of the Riemann-Roch theorem which we shall use in the next lecture. This is a result which enables one to compute the sheaf ω_D in some cases:

THEOREM 3: Let F be a non-singular projective surface. Then there is a canonical invertible sheaf Ω on F with the following property: Let $D \subset F$ be any effective divisor. Then D is a curve and

$$\omega_D \cong [\Omega \otimes \underline{o}_F(D)] \otimes \underline{o}_D .$$

Example: If $F = \mathbf{P}_2$, then, as is well-known, $\Omega = \underline{o}(-3)$. Then suppose $D \subset \mathbf{P}_2$ is a plane curve of degree d, i.e., $\underline{o}_{\mathbf{P}_2}(D) = \underline{o}(d)$. Then Theorem 3 tells us that

$$\omega_D \cong \underline{o}(d-3) \otimes \underline{o}_D .$$

For example, if $d = 3$, then $\omega_D \cong \underline{o}_D$, and for all invertible sheaves L on D, $H^i(D, L)$ and $H^{1-i}(D, L^{-1})$ are dual; in particular $H^1(D, \underline{o}_D)$ and $H^0(D, \underline{o}_D)$ are dual, hence

$$\chi(\underline{o}_D) = 0$$

$$p_a(D) = 1 .$$

Such curves are known as elliptic curves when D is non-singular.

LECTURE 12

THE OVER-ALL CLASSIFICATION OF CURVES ON SURFACES

We now turn our attention to geometry on a fixed projective and non-singular surface, F. On F we have divisors (Weil or Cartier, it makes no difference), and the group of divisor classes $\mathrm{Pic}(F)$. Among divisors, the effective divisors will be referred to simply as curves: these are now 1-dimensional closed subschemes, but they are not necessarily reduced <u>or</u> irreducible.

1° Let $D \subset F$ be a curve. Unlike the case of effective divisors on curves themselves, one cannot count the number of points in the support and call it the degree, since the support is positive dimensional. What we can do in the way of counting is this:

Let D_1, D_2 be two curves in F such that

$$\dim(\mathrm{Supp}(D_1) \cap \mathrm{Supp}(D_2)) = 0 .$$

Let $\{x_1, \ldots, x_n\} = \mathrm{Supp}(D_1) \cap \mathrm{Supp}(D_2)$.

Let f_i (resp. g_i) be a local equation for D_1 (resp. D_2) at x_i.

Define

$$(D_1 \cdot D_2) = \sum_{i=1}^{n} \dim_k[\underline{o}_{x_i}/(f_i, g_i)] .$$

This makes sense because the ideal (f_i, g_i) defines a subscheme of F at x_i which is set-theoretically the intersection $\mathrm{Supp}(D_1) \cap \mathrm{Supp}(D_2)$, i.e., which is $\{x_i\}$ itself. Therefore

$$(f_i, g_i) \supset m_{x_i}^N$$

for some N, and the dimension is finite.

This is the intersection number of D_1 and D_2, and it is easy to check that it is bilinear whenever defined. Like the degree in the geometry on curves, it depends only on the divisor classes, not the divisors:

<u>Proposition</u> 1: If $(D_1 \cdot D_2)$ is defined, then

$$(D_1 \cdot D_2) = \chi(\underline{o}_F) - \chi(\underline{o}_F(-D_1)) - \chi(\underline{o}_F(-D_2)) + \chi(\underline{o}_F(-D_1 - D_2)).$$

Proof: Consider the two complexes of sheaves

$$\underline{o}_F(-D_1) \to \underline{o}_F$$

and

$$\underline{o}_F(-D_2) \to \underline{o}_F .$$

Tensoring them, we get the complex

$$(*) \qquad \underline{o}_F(-D_1 - D_2) \to \underline{o}_F(-D_1) \oplus \underline{o}_F(-D_2) \to \underline{o}_F .$$

Since the original complexes are resolutions of $\underline{o}_{D_1}$ and $\underline{o}_{D_2}$ by locally free $\underline{o}_F$-modules, the cohomology of (*) consists in the sheaves

$$\mathrm{Tor}_i^{\underline{o}_F}(\underline{o}_{D_1}, \underline{o}_{D_2}) .$$

But if $x \in F$, and if f and g are local equations of D_1 and D_2 at x, then f and g are either one or both units, or f and g are an $\underline{o}_x$-sequence. In either case, the groups

$$\mathrm{Tor}_i^{\underline{o}_x}(\underline{o}_x/(f), \underline{o}_x/(g)) = (0), \qquad i > 0 .$$

Therefore (*) is a resolution of $\underline{o}_{D_1} \otimes \underline{o}_{D_2}$, which has stalk $\underline{o}_x/(f, g)$ at x. Therefore this sheaf is (0) except at $x_1, \ldots, x_n$, and at x_i it is isomorphic to

$$\underline{o}_{x_i}/(f_i, g_i) .$$

Therefore,

$$\begin{aligned}(D_1 \cdot D_2) &= \dim H^0(F, \underline{o}_{D_1} \otimes \underline{o}_{D_2}) \\ &= \chi(\underline{o}_{D_1} \otimes \underline{o}_{D_2}) \\ &= \chi(\underline{o}_F) - \chi[\underline{o}_F(-D_1) \oplus \underline{o}_F(-D_2)] + \chi(\underline{o}_F(-D_1 - D_2)) \\ &= \chi(\underline{o}_F) - \chi(\underline{o}_F(-D_1)) - \chi(\underline{o}_F(-D_2)) + \chi(\underline{o}_F(-D_1 - D_2)) .\end{aligned}$$

QED

This motivates:

Definition: Let L_1 and L_2 be any invertible sheaves on F.

$$(L_1 \cdot L_2) = \chi(\underline{o}_F) - \chi(L_1^{-1}) - \chi(L_2^{-1}) + \chi(L_1^{-1} \otimes L_2^{-1}) .$$

If D_1 and D_2 are any divisors on F, then

$$(D_1 \cdot D_2) = (\underline{o}_F(D_1) \cdot \underline{o}_F(D_2)) .$$

Proposition 2: (,) is a symmetric integral bilinear pairing, i.e.,

i) $(L_1 \cdot L_2) = (L_2 \cdot L_1)$

ii) $(L_1 \otimes L_1' \cdot L_2) = (L_1 \cdot L_2) + (L_1' \cdot L_2)$

iii) $(L_1^{-1} \cdot L_2) = -(L_1 \cdot L_2)$.

Proof: (i) is obvious, and (iii) follows from (ii) in virtue of the obvious fact:

$$(\underline{o}_F \cdot L) = 0.$$

In fact, I claim:

$$(\underline{o}_F(D) \cdot L) = \deg_D[L \otimes \underline{o}_D]$$

for any curve D on F. Use the sequences:

$$0 \to \underline{o}_F(-D) \to \underline{o}_F \to \underline{o}_D \to 0$$

and

$$0 \to L^{-1} \otimes \underline{o}_F(-D) \to L^{-1} \to (L \otimes \underline{o}_D)^{-1} \to 0 \quad .$$

Therefore,

$$\begin{aligned}(\underline{o}_F(D) \cdot L) &= [\chi(\underline{o}_F) - \chi(\underline{o}_F(-D))] - [\chi(L^{-1}) - \chi(L^{-1} \otimes \underline{o}_F(-D))] \\ &= \chi(\underline{o}_D) - \chi((L \otimes \underline{o}_D)^{-1}) \\ &= \deg_D[L \otimes \underline{o}_D] \ .\end{aligned}$$

Therefore, if L_2 admits a section, $(L_1 \cdot L_2)$ is linear in L_1, by the Riemann-Roch theorem (Theorem 1, Lecture 11).

Finally, let $\underline{o}(1)$ be a very ample invertible sheaf on F. If L is any invertible sheaf on F, then $L(n)$ has a section if n is large, by Serre's theorems. Now by writing the whole thing out one checks that the expression

$$(L_1 \cdot L_2) + (L_1' \cdot L_2) - (L_1 \otimes L_1' \cdot L_2)$$

is symmetric in the three variables L_1, L_1' and L_2. Since it is 0 when L_2 admits a section, it is also 0 when L_1' admits a section. Taking $L_1' = \underline{o}(n)$, this implies that

$$(L_1 \cdot L_2) = (L_1(n) \cdot L_2) - (\underline{o}(n) \cdot L_2) \ .$$

But both $\underline{o}(n)$ and $L_1(n)$ admits sections, hence the two terms on the right are linear in L_2. Therefore $(L_1 \cdot L_2)$ is linear in L_2.

QED

This bilinear form on Pic (F) takes the place of the degree homomorphism on Pic (X) for X a curve. It induces the following decomposition:

Definition: $\mathrm{Pic}^{\tau}(F)$ is the subgroup of Pic (F) consisting of those invertible sheaves L such that

$$(L \cdot L') = 0$$

all $L' \in \mathrm{Pic}\ (F)$.

Definition: $\mathrm{Num}\ (F) = \mathrm{Pic}\ (F)/\mathrm{Pic}^{\tau}(F)$.

By definition, Num (F) - the numerical divisor class group of F — is endowed with a non-degenerate symmetric integral pairing into **Z**. The fundamental result concerning Num (F), due to Severi and Néron, is that it is finitely generated as an abelian group; hence isomorphic to $\mathbf{Z}^{\rho}$, for some integer ρ, known as the base number of F. We will not need or prove this theorem (for the best proof, however, cf. LANG-NÉRON, Am. J. Math., 1959, Rational points of abelian varieties over function fields)

2° Although to understand the whole situation concerning the numerical characters of a divisor class $\{D\}$ one must look at its image in Num (F) or, equivalently, at the numbers $(\underline{o}_F(D) \cdot L)$, for all L, nonetheless for most purposes some of these numbers are more important and usually suffice:

Definition: If $\underline{o}(1)$ is a fixed very ample invertible sheaf on F, then relative to $\underline{o}(1)$ one defines:

$$\deg (L) = (L \cdot \underline{o}(1))$$

and

$$\deg (D) = \deg[\underline{o}_F(D)] = \deg_D[\underline{o}_D \otimes \underline{o}(1)] .$$

Incidentally, if D is effective, then $\deg (D) > 0$: let $\underline{o}(1)$ on F be induced by:

$$i: F \hookrightarrow P_n .$$

Let $H \subset P_n$ be a hyperplane not containing any of the points $i(x)$, x a generic point of Supp (D). Then the curve $H' = i^*(H)$ is defined and

$$\dim \{\mathrm{Supp}\,(D) \cap \mathrm{Supp}\,(H')\} = 0 .$$

Therefore,

$$\begin{aligned}\deg (D) &= (\underline{o}_F(D) \cdot \underline{o}_F(H')) \\ &= (D \cdot H') \\ &\geq 0 .\end{aligned}$$

But suppose $\deg (D) = 0$; then $\mathrm{Supp}\,(D) \cap \mathrm{Supp}\,(H') = \emptyset$. To prevent this, choose a closed point $y \in \mathrm{Supp}\,(D)$ and choose the hyperplane H such that $i(y) \in H$ while $i(x)$ is still not in H for generic points $x \in \mathrm{Supp}\,(D)$. This is certainly possible, and, therefore $\deg (D) > 0$.

Returning to an arbitrary invertible sheaf L on F, the other number of great importance is its Euler characteristic. This number is given by an intersection product too. To derive this, use the third part of the Riemann-Roch theorem on curves.

Proposition 3: Let L be an invertible sheaf on F, and let Ω be the canonical invertible sheaf on F given by Theorem 3, Lecture 11. Then

$$\chi(L) = \tfrac{1}{2}(L \cdot L \otimes \Omega^{-1}) + \chi(\underline{o}_F) .$$

Proof: The formula states:

$$
\begin{aligned}
2(\chi(L) - \chi(\underline{o}_F)) &= (L \cdot L \otimes \Omega^{-1}) \\
&= -(L^{-1} \cdot L \otimes \Omega^{-1}) \\
&= -\chi(\underline{o}_F) + \chi(L) + \chi(L^{-1} \otimes \Omega) - \chi(\Omega)
\end{aligned}
$$

or

$$(\#) \qquad \chi(L) - \chi(\underline{o}_F) - \chi(\Omega \otimes L^{-1}) + \chi(\Omega) = 0 .$$

If L^{-1} has a section, then $L \cong \underline{o}_F(-D)$ for some curve D. Then use the exact sequences:

$$0 \to L \to \underline{o}_F \to \underline{o}_D \to 0$$

and

$$0 \to \Omega \to \Omega \otimes L^{-1} \to \omega_D \to 0$$

(cf. Theorem 3, Lecture 11). By Theorem 2, Lecture 11, $\chi(\omega_D) + \chi(\underline{o}_D) = 0$, hence (#) follows whenever L^{-1} has a section.

Finally, let $\underline{o}(1)$ be a very ample invertible sheaf on F. If M is any invertible sheaf on F, then $M^{-1}(n)$ and $\underline{o}(n)$ both have sections if n is large, by Serre's theorems. Now a simple computation shows that the expression on the left in (#) is linear in L. Namely:

$$
\begin{aligned}
&[\chi(L \otimes M) - \chi(\underline{o}_F) - \chi(\Omega \otimes L^{-1} \otimes M^{-1}) + \chi(\Omega)] \\
&\quad - [\chi(L) - \chi(\underline{o}_F) - \chi(\Omega \otimes L^{-1}) + \chi(\Omega)] \\
&\quad - [\chi(M) - \chi(\underline{o}_F) - \chi(\Omega \otimes M^{-1}) + \chi(\Omega)] \\
&= \{\chi(\underline{o}_F) - \chi(L) - \chi(\Omega \otimes L^{-1} \otimes M^{-1}) + \chi(\Omega \otimes M^{-1})\} \\
&+ \{\chi(\underline{o}_F) - \chi(M) - \chi(\Omega \otimes L^{-1} \otimes M^{-1}) + \chi(\Omega \otimes L^{-1})\} \\
&- \{\chi(\underline{o}_F) - \chi(L \otimes M) - \chi(\Omega \otimes L^{-1} \otimes M^{-1}) + \chi(\Omega)\} \\
&= (L^{-1} \cdot \Omega^{-1} \otimes L \otimes M) + (M^{-1} \cdot \Omega^{-1} \otimes L \otimes M) \\
&- (L^{-1} \otimes M^{-1} \cdot \Omega^{-1} \otimes L \otimes M) \\
&= 0 .
\end{aligned}
$$

But then the expression in (#) is 0 for $L = M(-n)$ and for $L = \underline{o}(-n)$ by the first part of the proof. Therefore it is 0 for $L = M$.

QED

This result is the weakest version of the Riemann-Roch theorem on F. As one consequence of this result, we see that the only really important numerical characters of an invertible sheaf L are

$$\left.\begin{aligned} \deg(L) &= (L \cdot \underline{o}(1)) \\ (L^2) &= (L \cdot L) \\ &(L \cdot \Omega) \end{aligned}\right\}$$

and

3° So far we have studied the discrete aspects of Pic (F), and hence the discrete aspects of the set of curves on F. To get at the existence questions of Lecture 2, we shall look at the continuous part of these two sets. The "glueing" which gives continuity must come from the concept of families of invertible sheaves and families of curves. We make the following definitions:

Definition. Let S be a scheme (algebraic /k). A family of curves on F, over S, is a relative effective Cartier divisor $\mathfrak{D} \subset F \times S$, over S. A family of invertible sheaves on F, over S, is an invertible sheaf L on $F \times S$: except that two invertible sheaves L_1, L_2 will be said to define the same family of invertible sheaves if there is an invertible sheaf M on S such that:

$$L_1 \cong L_2 \otimes p_2^*(M) \quad .$$

How does the concept of a family really provide the glueing? This comes about because the collection of families forms a functor:

a) $\underline{\text{Curves}}_F(S)$ = set of families of curves on F over S

and

b) $\underline{\text{Pic}}_F(S)$ = set of families of invertible sheaves on F over S.

Given $T \xrightarrow{g} S$, one obtains:

$$F \times T \xrightarrow{h} F \times S \; ;$$

hence for $\mathfrak{D} \subset F \times S$ (resp. L on $F \times S$), one obtains $h^*(\mathfrak{D}) \subset F \times T$ (resp. $h^*(L)$ on $F \times T$). This is a map

a) $\underline{\text{Curves}}_F(S) \xrightarrow{g^*} \underline{\text{Curves}}_F(T)$

and

b) $\underline{\text{Pic}}_F(S) \xrightarrow{g^*} \underline{\text{Pic}}_F(T)$.

The glueing is now equivalent to the problem of representing these functors: to represent these functors is the same as to find a universal family of curves or invertible sheaves. And if you find such a family, say over S, then the set of k-rational points of S will be canonically isomorphic to the set of curves on F, or to the set Pic (F); i.e., you have put these sets together into whole schemes. Notice also that we have a morphism of functors:

$$\underline{\text{Curves}}_F \xrightarrow{\Phi} \underline{\text{Pic}}_F$$

which maps $\mathfrak{D} \subset F \times S$ to the invertible sheaf $\underline{o}_{F\times S}(\mathfrak{D})$. Consequently if C (resp. P) were schemes representing these two functors, one would automatically get a morphism of schemes,

$$C \xrightarrow{\varphi} P$$

which, on k-rational points, restricts to the obvious map from the set of curves on F to the set $\underline{\text{Pic}}$ (F).

In terms of this glueing, we can say precisely why the numerical invariants of 1°, 2° are discrete. Say L_1, L_2 are two invertible sheaves on $F \times S$. For each closed point $s \in S$, they induce sheaves $L_{1,s}$ and $L_{2,s}$ on the fibre F, and we can compute $(L_{1,s} \cdot L_{2,s})$: this number is constant on each connected component of S ! [Since $(L_{1,s} \cdot L_{2,s})$ is a sum of Euler characteristics and these are values of Hilbert polynomials, this follows from Corollary 3, Lecture 7.] In other words, given any family of invertible sheaves over a connected base S, the image of each sheaf L_s in Num (F) is the same. Therefore, if an object P represents the functor $\underline{\mathrm{Pic}}_F$, for each element of Num (F), the set of invertible sheaves inducing this element would form an open and closed set of P. The natural thing to do is to break up the functors $\underline{\mathrm{Pic}}_F$ and $\underline{\mathrm{Curve}}_F$ accordingly into manageable pieces:

__Definition__: Let $\xi \in$ Num (F). For all schemes S, let $\underline{\mathrm{Pic}}_F^\xi(S)$ be the subset of $\underline{\mathrm{Pic}}_F(S)$ consisting of those L on $F \times S$ such that for all closed points $s \in S$, if L_s is the induced sheaf on F over s, then L_s has numerical class ξ. Moreover, let $\underline{\mathrm{Curves}}_F^\xi(S)$ be the subset of $\underline{\mathrm{Curves}}_F(S)$ mapped by Φ into $\underline{\mathrm{Pic}}_F^\xi(S)$. Both form subfunctors denoted Curves_F^ξ and $\underline{\mathrm{Pic}}_F^\xi$.

The principal results at which we are aiming are:

FIRST CONSTRUCTION THEOREM: For all ξ, $\underline{\mathrm{Curves}}_F^\xi$ is isomorphic to a functor $h_{C(\xi)}$, where $C(\xi)$ is a projective scheme.

SECOND CONSTRUCTION THEOREM: For all ξ, $\underline{\mathrm{Pic}}_F^\xi$ is isomorphic to a functor $h_{P(\xi)}$, where $P(\xi)$ is a projective scheme.

As a corollary, it follows readily that the full functors $\underline{\mathrm{Curves}}_F$ and $\underline{\mathrm{Pic}}_F$ are represented by (non-algebraic) schemes which are the disjoint unions:

$$\coprod_\xi C(\xi) \quad \text{and} \quad \coprod_\xi P(\xi) \ .$$

LECTURE 13

LINEAR SYSTEMS AND EXAMPLES

Before looking at the general problem of constructing $C(\xi)$ and $P(\xi)$, we want to describe some special cases in which the answer is very simple and then to show how some of the Examples of Lecture 1 fall in this category, hence can now be treated rigorously.

1° We start with a case in which the group $\operatorname{Pic}(F)$ and hence the group $\operatorname{Num}(F)$ is particularly simple:

Assume i) $H \subset F$ is an irreducible curve,

ii) $F - H$ is affine,

iii) $\Gamma(F - H, \underline{o}_F)$ is a unique factorization domain.

<u>Proposition 1</u>: Then $\operatorname{Pic}(F)$ is an infinite cyclic group generated by the image h of H; and

$$\operatorname{Pic}(F) \cong \operatorname{Num}(F).$$

<u>Proof</u>: We must show that any divisor D on F is linearly equivalent to nH for some integer n. Since divisors are Weil divisors, every divisor is the difference of two effective divisors and we may as well assume that D is effective. Let the closed subscheme $D \cap (F - H)$ of $F - H$ correspond to the ideal

$$\mathfrak{A} \subset R = \Gamma(F - H, \underline{o}_F) .$$

Since $\mathfrak{A}$ induces a principal ideal in each localization $R_{\wp}$ of R, it follows that all prime ideals associated to $\mathfrak{A}$ are minimal; hence, since R is a UFD, $\mathfrak{A}$ itself is principal. Let $\mathfrak{A} = (f)$. Then the divisor $D - (f)$ has neither zeroes nor poles in $F - H$, i.e., $\operatorname{Supp}[D - (f)] \subset H$. This menas that

$$D - (f) = nH, \text{ some } n \in \mathbb{Z} ,$$

hence $D \equiv nH$. Therefore h generates $\operatorname{Pic}(F)$, and hence $\operatorname{Num}(F)$. It remains to check that $\operatorname{Num}(F)$ is infinite cyclic—for then so is $\operatorname{Pic}(F)$ and these two groups are isomorphic. But since F is projective, the divisor nH is very ample for some n (i.e., $\underline{o}_F(nH)$ is of the form $\underline{o}(1)$).

Therefore, as remarked in Lecture 12,

$$\begin{aligned} n(H \cdot H) &= (\underline{o}_F(H) \cdot \underline{o}_F(nH)) \\ &= (\underline{o}_F(H) \cdot \underline{o}(1)) \\ &= \deg H \\ &> 0 \end{aligned}$$

and therefore the image of h in Num (F) has infinite order.

QED

Clearly this result applies to $\mathbf{P}_2$, since if H is a hyperplane,

$$\Gamma(\mathbf{P}_2 - H, \underline{o}_{\mathbf{P}_2}) \cong k[X, Y] .$$

Therefore, all curves D in $\mathbf{P}_2$ have some degree d, and $D \equiv dH$, i.e., $\underline{o}_{\mathbf{P}_2}(D) \cong \underline{o}_{\mathbf{P}_2}(d)$. Since $H^0(\mathbf{P}_2, \underline{o}_{\mathbf{P}_2}(d))$ is spanned by homogeneous forms in the homogeneous coordinates X_0, X_1, X_2 of degree d, it follows that all curves on $\mathbf{P}_2$ are of the type we expect.

Incidently, the Proposition is valid in any dimension, so it can be applied to various Grassmannians, Hyperquadrics, etc, (also to hypersurfaces of some types, cf. ANDREOTTI, SALMON, Monatshefte fur Math., 61, 1957, p. 97).

2° In cases where the Picard group is simple, the set of curves is also fairly simple. Actually, what is always simple are the fibres in the set of curves over the Picard group, i.e., the set of curves linearly equivalent to a fixed curve. However, to state their structure properly, again we have to find the glue to put these "linear systems" of curves together. What is required is the fibre of the morphism Φ from the functor $\underline{\text{Curves}}_F$ to the functor $\underline{\text{Pic}}_F$.

Quite generally, Grothendieck has defined the fibres of a morphism of functors. Let F, G be contravariant functors from a category $\mathcal{C}$ to (Sets). Let $\Phi: F \to G$ be a morphism. Let S be an object in $\mathcal{C}$, and let $\alpha \in G(S)$: we shall define the fibre of Φ over α. It is to be a functor too, but not from $\mathcal{C}$ to (Sets). It is a functor from the category $\mathcal{C}/S$ of objects over S [i.e., an object is a morphism $T \xrightarrow{f} S$, and a morphism is a commutative diagram

$$\begin{array}{ccc} T_1 & \xrightarrow{g} & T_2 \\ & {\scriptstyle f_1}\searrow \quad \swarrow{\scriptstyle f_2} & \\ & S & \end{array}$$]

to the category (Sets). Call it Φ^α:

$$\Phi^\alpha(T \xrightarrow{f} S) = \{\beta \in F(T) \mid \Phi(\beta) = f^*(\alpha) \text{ in } G(T)\} .$$

The rest of the definition is clear.

In our case, $\mathcal{C}$ is the category of algebraic schemes over k; and $\alpha \in G(\mathrm{Spec}\ (k))$, i.e., α would be a closed point of the object representing G. Then Φ^{α} is again a functor on the category of algebraic schemes over k because $\mathrm{Spec}\ (k)$ is the final object in this category. The key point is this: <u>If F and G are represented by schemes X and Y, then Φ is induced by a morphism $\varphi: X \to Y$, α is a closed point of Y and Φ^{α} is represented by the actual fibre $\varphi^{-1}(\alpha)$.</u>

(Proof: immediate.)

In the case of $\underline{\mathrm{Curves}}_F$ and $\underline{\mathrm{Pic}}_F$, the fibre functor is:

<u>Definition:</u> Let L be an invertible sheaf on F. Let

$$\underline{\mathrm{Lin}}\ \underline{\mathrm{Sys}}_L(S) = \{\mathfrak{D} \subset F \times S \mid \mathfrak{D} \text{ a relative effective Cartier divisor over } S \text{ such that } \underline{o}_{F\times S}(\mathfrak{D}) \cong p_1^*(L) \otimes p_2^*(K) \text{ for some invertible sheaf } K \text{ on } S\}\ .$$

Via the usual maps, this is a contravariant functor in S.

In Lecture 1 we gave heuristic reasons for describing $\underline{\mathrm{Lin}}\ \underline{\mathrm{Sys}}_L$ as a projective space. The full result can now be proven:

<u>Proposition 2:</u> Let L be any invertible sheaf on F. Let $N = \dim H^0(F, L)$. Then

$$\underline{\mathrm{Lin}}\ \underline{\mathrm{Sys}}_L \cong h_{P_{N-1}}\ .$$

<u>Proof:</u> Suppose $D \subset F \times S$ is an element of $\underline{\mathrm{Lin}}\ \underline{\mathrm{Sys}}_L\ (S)$: then $\underline{o}_{F\times S}(D) \cong p_1^*(L) \otimes p_2^*(K)$. In other words, D is determined by an invertible sheaf K on S, and a section:

$$s \in H^0(F \times S,\ p_1^*(L) \otimes p_2^*(K))$$

[i.e., the image of $1 \in \Gamma(F \times S, \underline{o}_{F\times S}(D))$]. Moreover, since the Cartier divisor $s = 0$ is relative over S, it must happen that $s(x) \neq 0$ for all x in $A(p_2^{-1}(p_2(x)))$. Now if $y \in S$, and $K = \mathcal{K}(y)$, then the fibre $p_2^{-1}(y)$ over y is just $F \underset{\mathrm{spec}(k)}{\times} \mathrm{Spec}\ (K)$: this is reduced and irreducible since F is a variety, hence its only associated point is its generic point. Therefore the condition on s is just that $s \not\equiv 0$ on any fibre $p_2^{-1}(y)$ of p_2.

Now suppose K_1 and s_1 determine the same D as K_2 and s_2: I claim that there is an isomorphism of K_1 and K_2 under which the sections s_1 and s_2 correspond. Now we have isomorphisms:

$$p_1^*(L) \otimes p_2^*(K_1) \cong \underline{o}_{F\times S}(D) \cong p_1^*(L) \otimes p_2^*(K_2)\ .$$

Let $\varepsilon: S \to F \times S$ be a section of p_2 gotten by mapping S to $\{x\} \times S$ $F \times S$, for some closed point $x \in F$. Then:

$$K_1 \cong \varepsilon^*\{p_1^*(L) \otimes p_2^*(K_1)\} \cong \varepsilon^*\{p_1^*(L) \otimes p_2^*(K_2)\} \cong K_2\ .$$

Therefore, we may as well assume $K_1 = K_2$. Now if the sections s_1, s_2 are not equal, they differ by an element

$$\alpha \in H^0(F \times S, \underline{o}^*_{F\times S})$$

since they define the same Cartier divisor. But

$$\begin{aligned} H^0(F \times S, \underline{o}^*_{F\times S}) &= H^0(S, p_{2,*}(\underline{o}^*_{F\times S})) \\ &= H^0(S, \underline{o}^*_S) \;. \end{aligned}$$

Therefore, if we modify the identification of K_1 and K_2 by this scalar, we can assume $s_1 = s_2$. Thus

$$\left\{ \begin{array}{l} \text{set of families of curves} \\ D \subset F \times S \text{ in } \underline{\text{Lin}}\ \underline{\text{Sys}}_L \end{array} \right\} \cong \left\{ \begin{array}{l} \text{set of invertible sheaves } K \text{ on } S, \\ \text{and sections of } p_1^*(L) \otimes p_2^*(K) \text{ not} \\ \text{zero on any fibre of } p_2 \text{ — up to iso-} \\ \text{morphism.} \end{array} \right\}$$

Now recall that:

$$H^0(F \times S, p_1^*(L) \otimes p_2^*(K)) = H^0(F, L) \otimes H^0(S, K)$$

by the Künneth formula (Lecture 11, (IV.)). Fix a basis $e_1, \ldots, e_N$ of $H^0(F, L)$. Then sections of $p_1^*(L) \otimes p_2^*(K)$ are of the form

$$s = \sum_{i=1}^{N} e_i \otimes s_i \;,$$

for $s_i \in H^0(S, K)$. Moreover $s \equiv 0$ on $p_2^{-1}(y)$ if and only if $s_i(y) = 0$ for all i. Therefore:

$$\left\{ \begin{array}{l} \text{set of invertible sheaves} \\ K \text{ on } S, \text{ and sections of} \\ p_1^*(L) \otimes p_2^*(K) \text{ not zero on} \\ \text{any fibre of } p_2 \text{ —up to} \\ \text{isomorphism} \end{array} \right\} \cong \left\{ \begin{array}{l} \text{set of invertible sheaves } K \text{ on} \\ S, \text{ and } N \text{ sections } s_1, \ldots, s_N \\ \text{of } K \text{ not all simultaneously} \\ \text{zero at any } y \in S \text{ —up to iso-} \\ \text{morphism.} \end{array} \right\}$$

But the latter is exactly the set of S-valued points of $\mathbf{P}_{N-1}$. This sets up an isomorphism of the functors $\underline{\text{Lin}}\ \underline{\text{Sys}}_L$ and $h_{\mathbf{P}_{N-1}}$.

QED

Looking more closely at the proof of this Proposition, one can say that the space $\mathbf{P}_{N-1}$ representing $\underline{\text{Lin}}\ \underline{\text{Sys}}_L$ is not just any projective space: if it is identified <u>canonically</u>, it is the projective space

$$\mathbf{P}[\widehat{H^0(F, L)}] \;,$$

(where $\hat{V}$ is the dual vector space to V).

3° It would seem as if we were now in a position to describe $C(\xi)$ and $P(\xi)$ completely in simple cases: for $\mathbf{P}_2$, Pic ($\mathbf{P}_2$) is very simple and the fibres of Φ are always easy. But there is one possibility still to be checked: even the discrete set of points Pic ($\mathbf{P}_2$) = $\mathbf{Z}$ could be endowed with nontrivial scheme structure, i.e., nilpotents in its structure sheaf. In fact, this occurs for some surfaces, and even under the assumptions of 1° (as far as I know) an additional hypothesis is needed to prevent this situation. Also, in Lecture 1, we saw quite a few other cases where the only families of curves were linear systems, so that Pic (F) was a discrete set. We need a direct way of checking when this will happen:

Proposition 3: Suppose $H^1(F, \underline{o}_F) = (0)$. Let S be any connected algebraic scheme, and let $\mathfrak{L}$ be an invertible sheaf on $F \times S$. Then there are invertible sheaves L on F and K on S such that:

$$\mathfrak{L} \cong p_1^*(L) \otimes p_2^*(K) .$$

Proof: For all closed points $s \in S$, $\mathfrak{L}$ induces an invertible sheaf L_s on the fibre $p_2^{-1}(s) = F$. Let

$$M_s = \mathfrak{L} \otimes p_1^*(L_s^{-1}) .$$

Look at the cohomology of M_s with respect to p_2.

a) the induced sheaf $M_s \otimes \mathfrak{K}(s)$ on the fibre $p_2^{-1}(s)$ is isomorphic to $\underline{o}_F$ by the very definition of M_s;

b) therefore, by the key hypothesis of the Proposition,

$$H^1(p_2^{-1}(s), M_s \otimes \mathfrak{K}(s)) = (0) .$$

Using Corollary 1 in 3°, Lecture 7, all sections in $H^0(p_2^{-1}(s), M_s \otimes \mathfrak{K}(s))$ lift to sections of $p_{2,*}(M_s)$ in some neighborhood of s.

c) But as $M_s \otimes \mathfrak{K}(s) = \underline{o}_F$, the section 1 of $\underline{o}_F$ lifts to a section:

$$\alpha \in \Gamma(U, p_{2,*}(M_s)) = H^0(F \times U, M_s) .$$

d) Then α defines a homomorphism:

$$p_1^*(L_s) \xrightarrow{\varphi} \mathfrak{L}$$

in $F \times U$. Moreover, since α comes from 1 in $p_2^{-1}(s)$, φ is an isomorphism of the induced sheaves L_s and $\mathfrak{L} \otimes \mathfrak{K}(s)$ on the fibre $p_2^{-1}(s)$. Therefore φ is an isomorphism of $p_1^*(L_s)$ and $\mathfrak{L}$ at all points over s, hence φ is an isomorphism in an open neighborhood W of $p_2^{-1}(s)$. Since p_2: $F \times S \to S$ is topologically closed, there is an open neighborhood $U_s \subset U$ of s such that $W \supset F \times U_s$. This proves that $p_1^*(L_s)$ and $\mathfrak{L}$ are isomorphic to $F \times U_s$.

e) Therefore if $s' \in U_s$, $L_{s'}$ and L_s are isomorphic. Since S is connected, this implies that all the sheaves L_s are isomorphic. Call this sheaf L. Then we have an open covering U_i of S such that $p_1^*(L)$ and $\mathfrak{L}$ are isomorphic in each open set $F \times U_i$.

f) Fix isomorphisms

$$\psi_i: \ p_1^*(L) \xrightarrow{\sim} \mathfrak{L}$$

in $F \times U_i$. Then in $F \times (U_i \cap U_j)$, $\psi_j^{-1} \circ \psi_i$ is an automorphism of $p_1^*(L)$.

This is given by multiplication by a unit:

$$\sigma_{ij} \in \Gamma(F \times (U_i \cap U_j), \underline{o}^*_{F\times S}) \cong \Gamma(U_i \cap U_j, \underline{o}^*_S)$$

(cf. Lecture 11,IV). Then $\{\sigma_{ij}\}$ is a 1-Czech-co-cycle on $\underline{S}$ for the covering $\{U_i\}$. Let this co-cycle be the transition functions for an invertible sheaf K on S. Then it follows from our construction that $\mathfrak{L}$ is isomorphic globally to $p_1^*(L) \otimes p_2^*(K)$.

QED

This result is closely related to the see-saw principle of LANG (cf. his Abelian Varieties).

Corollary: If $H^1(F, \underline{o}_F) = (0)$, then $\underline{\mathrm{Pic}}_F$ is represented by the disjoint union of a (infinite) discrete set of points, i.e., of Spec (k)'s. Therefore $\underline{\mathrm{Curves}}_F$ is represented by the disjoint union of projective spaces (of various dimensions).

This completes our justification of our description of curves on $\mathbf{P}_2$. Perhaps to add the last point, we should compute:

$$(\underline{o}(n) \cdot \underline{o}(m)) = n \cdot m \quad .$$

[immediate by bilinearity, and the check:

$$(o(1) \cdot \underline{o}(1)) = (H_1 \cdot H_2) = 1$$

for two distinct lines H_1, H_2 in $\mathbf{P}_2$].

Exercise: Write down explicitly the universal families of curves on $\mathbf{P}_2$.

Further Examples: Without proofs, we want to supplement Examples 2 and 5 of Lecture 1 by relating the results there to our present theory. Both of these surfaces are "birational" to $\mathbf{P}_2$, i.e., are isomorphic to $\mathbf{P}_2$ on open dense subsets. In fact, it follows from this that

$$H^1(F, \underline{o}_F) = H^2(F, \underline{o}_F) = (0)$$

in both these cases. Therefore both fall under the Corollary just given. Now, in the case $F = \mathbf{P}_1 \times \mathbf{P}_1$, then

$$\mathrm{Pic}\ (F) \cong \mathrm{Num}\ (F) \cong \mathbf{Z} \otimes \mathbf{Z} \ .$$

In fact, a basis is given by the two sheaves

$$L_1 = p_1^*(\underline{o}(1)) \quad \text{and } L_2 = p_2^*(\underline{o}^*1))$$

and the degrees d and e of a divisor D described before are just the d and e defined by:

$$\underline{o}_F(D) \cong L_1^e \otimes L_2^d \ .$$

The pairing is given by

$$(L_1 \cdot L_1) = 0$$
$$(L_1 \cdot L_2) = 1$$
$$(L_2 \cdot L_2) = 0 \ .$$

Now in case where F is obtained by blowing up two points in P_2,

$$\text{Pic }(F) \cong \text{Num }(F) \cong \mathbf{Z} \otimes \mathbf{Z} \otimes \mathbf{Z} \ .$$

In fact, a basis is given by the three sheaves

$$M_1 = \underline{o}_F(E_1), \quad M_2 = \underline{o}_F(E_2), \quad L = \underline{o}_F(D) \quad .$$

The pairing is given by:

$$\begin{bmatrix} (M_1 \cdot M_1) = -1 & (M_1 \cdot M_2) = 0 & (M_1 \cdot L) = 1 \\ (M_2 \cdot M_1) = 0 & (M_2 \cdot M_2) = -1 & (M_2 \cdot L) = 1 \\ (L \cdot M_1) = 1 & (L \cdot M_2) = 1 & (L \cdot L) = -1 \end{bmatrix}$$

LECTURE 14

SOME VANISHING THEOREMS

Some of the deepest results in algebraic geometry concern the problem of giving criteria for the higher cohomology groups of a sheaf to be 0. The pivotal role played by these results is due to the fact that the Euler characteristic of a coherent sheaf on some variety is generally very computable: either directly, or by use of the very powerful Hirzebruch-Grothendieck form of the Riemann-Roch Theorem; on the other hand, it is usually the group of sections of such sheaves which has geometric interest and direct significance. Therefore, whenever one can prove that the higher cohomology is 0, one should expect many corollaries.

A first theorem of this type was proven in Lecture 11. The general problem was formulated by the Italians: it was known as the problem of <u>postulation</u> (i.e., when does the dimension of something turn out to equal the number which one had postulated!?). Picard proved by analytic methods a very famous result of this kind (the theorem of the regularity of the adjoint, cf ZARISKI's book on surfaces); this result was greatly extended by KODAIRA in one of his most famous papers (Proc. Natl. Acad. Sci., 1953, p. 1268: <u>A differential-geometric method in the theory of analytic stacks</u>), and today it is known as Kodaira's vanishing theorem. Another result in this direction is Serre's duality theorem (vastly extended by Grothendieck): this is the direct descendent of Roch's result and it tells, on an n-dimensional non-singular variety, how to compute an H^i by means of an H^{n-i}, which at least cuts the problem in half.

We shall prove here (with the help of techniques developed and used by Nakai, Matsusaka and Kleiman) only a weak vanishing theorem, but one which is <u>uniformly</u> applicable to a large class of sheaves. Let $\mathcal{F}$ be a coherent sheaf on $\mathbf{P}_n$:

<u>Definition:</u> $\mathcal{F}$ is <u>m-regular</u> if $H^i(\mathbf{P}_n, \mathcal{F}(m-i)) = (0)$ for all $i > 0$.

This apparently silly definition reveals itself as follows:

<u>Proposition:</u> (Castelnuovo) Let $\mathcal{F}$ be an m-regular coherent sheaf on $\mathbf{P}_n$. Then

a) $H^0(\mathbf{P}_n, \mathcal{F}(k))$ is spanned by

$$H^0(\mathbf{P}_n, \mathcal{F}(k-1)) \otimes H^0(\mathbf{P}_n, \underline{o}(1)) \quad \text{if} \quad k > m;$$

b) $H^i(\mathbf{P}_n, \mathcal{F}(k)) = (0)$ whenever $i > 0$, $k + i \geq m$.

Hence a') $\mathcal{F}(k)$ is generated as $\underline{o}_{\mathbf{P}_n}$-module by its global sections if $k \geq m$.

Proof: We use induction on n: for $n = 0$, the result is obvious. In general, given $\mathcal{F}$, choose a hyperplane H not containing any of the points in the finite set $A(\mathcal{F})$. Tensor the exact sequence:

$$0 \to \underline{o}_{\mathbf{P}_n}(-H) \to \underline{o}_{\mathbf{P}_n} \to \underline{o}_H \to 0$$

$$\underline{o}_{\mathbf{P}_n}(-H) \cong \underline{o}_{\mathbf{P}_n}(-1)$$

with $\mathcal{F}(k)$. For all $x \in \mathbf{P}_n$, if f is a local equation for H at x, then multiplication by f is injective in $\mathcal{F}_x$. since by construction, f is a unit at all associated primes of $\mathcal{F}_x$. Therefore the resulting sequence:

$$(*)_k \qquad 0 \to \mathcal{F}(k-1) \to \mathcal{F}(k) \to \underbrace{(\mathcal{F} \otimes \underline{o}_H)(k)}_{\mathcal{F}_H(k)} \to 0$$

is exact. In particular, we get:

$$H^i(\mathcal{F}(m-i)) \to H^i(\mathcal{F}_H(m-i)) \to H^{i+1}(\mathcal{F}(m-i-1)) .$$

This implies that if $\mathcal{F}$ is m-regular, the sheaf $\mathcal{F}_H$ on H is m-regular. Since $H \cong \mathbf{P}_{n-1}$, we use the induction hypothesis to obtain a) and b) for $\mathcal{F}_H$. In particular, use

$$H^{i+1}(\mathcal{F}(m-i-1)) \to H^{i+1}(\mathcal{F}(m-i)) \to H^{i+1}(\mathcal{F}_H(m-i)) .$$

If $i \geq 0$, by b) for $\mathcal{F}_H$, the last group is (0); by m-regularity the first group is (0). Therefore, the middle group is (0) and $\mathcal{F}$ is (m+1)-regular. Continuing in this way we prove b) for $\mathcal{F}$.

To get a), look at the diagram:

$$\begin{array}{ccc} & H^0(\mathcal{F}(k-1)) \otimes H^0(\underline{o}_{\mathbf{P}_n}(1)) & \xrightarrow{\sigma} & H^0(\mathcal{F}_H(k-1)) \otimes H^0(\underline{o}_H(1)) \\ & \downarrow \mu & & \downarrow \tau \\ H^0(\mathcal{F}(k-1)) \to & H^0(\mathcal{F}(k)) & \xrightarrow{\nu} & H^0(\mathcal{F}_H(k)) \end{array}$$

Note that σ is surjective if $k > m$ because $H^1(\mathcal{F}(k-2)) = (0)$. Moreover, τ is surjective if $k > m$ by conclusion a) for $\mathcal{F}_H$. Therefore, $\nu(\operatorname{Im} \mu)$ is the whole of $H^0(\mathcal{F}_H(k))$, i.e., $H^0(\mathcal{F}(k))$ is spanned by $\operatorname{Im}(\mu)$ and by $H^0(\mathcal{F}(k-1))$. But let $h \in H^0(\mathbf{P}_n, \underline{o}_{\mathbf{P}_n}(1))$ be the global equation of H. Then the image of $H^0(\mathcal{F}(k-1))$ in $H^0(\mathcal{F}(k))$ is more precisely $h \otimes H^0(\mathcal{F}(k-1))$. In other words, this is part of $\operatorname{Im} \mu$ too. Therefore μ is surjective and a) is proven for $\mathcal{F}$.

Now by Serre's theorem, we know that $\mathcal{F}(k)$ is generated by its sections provided that k is large enough. Putting this together with a) implies that $H^0(\mathcal{F}(m)) \otimes H^0(\underline{o}_{P_n}(k-m))$ generates the sheaf $\mathcal{F}(k)$ of $\underline{o}_{P_n}$-modules if $k \gg 0$. But for every $x \in P_n$, fix an isomorphism of $\underline{o}_{P_n}(1)$ and $\underline{o}_{P_n}$ at x: this identifies $\underline{o}_{P_n}(k-m)$ with $\underline{o}_{P_n}$ at x, and $\mathcal{F}(k)$ with $\mathcal{F}(m)$ at x. Then $H^0(\underline{o}_{P_n}(k-m))$ becomes just a vector space of elements of the local ring $\underline{o}_x$, and the statement simply says that $H^0(\mathcal{F}(m)) \otimes \underline{o}_x$ generates the stalk $\mathcal{F}(m)_x$, i.e., $\mathcal{F}(m)$ is generated by its global sections.

QED

Our main result is:

THEOREM: For all n, there is a polynomial $F_n(x_0, \ldots, x_n)$ such that for all coherent sheaves of ideals $\mathcal{I}$ on P_n, if $a_0, a_1, \ldots, a_n$ are defined by:

$$\chi(\mathcal{I}(m)) = \sum_{i=0}^{n} a_i \binom{m}{i} ,$$

then $\mathcal{I}$ is $F_n(a_0, a_1, \ldots, a_n)$-regular.

<u>Proof</u>: Again we use induction on n since for $n = 0$ the result is obvious. Given $\mathcal{I}$, let $Z \subset P_n$ be the corresponding subscheme; choose a hyperplane H such that H is disjoint from $A(\underline{o}_Z)$. As above, we get the exact sequence:

$$(*)_m \qquad 0 \rightarrow \mathcal{I}(m) \xrightarrow{\otimes h} \mathcal{I}(m+1) \rightarrow \underbrace{(\mathcal{I} \otimes \underline{o}_H)}_{\mathcal{I}_H}(m+1) \rightarrow 0$$

which is injective on the left since multiplication by a local equation for H is injective in the sheaf $\mathcal{I}$, as it is a subsheaf of $\underline{o}_{P_n}$. On the other hand, $\mathcal{I}_H$ is a sheaf of ideals on H: let $x \in P_n$ and let f be a local equation for H at x. Then

$$0 \rightarrow \mathcal{I}_x \rightarrow \underline{o}_{x,P_n} \rightarrow \underline{o}_{x,Z} \rightarrow 0$$

gives:

$$\mathrm{Tor}_1(\underline{o}/f\cdot\underline{o}_x, \underline{o}_{x,Z}) \rightarrow (\mathcal{I}_H)_x \rightarrow \underline{o}_{x,H}$$

by tensoring with $\underline{o}_x/f\cdot\underline{o}_x = \underline{o}_{x,H}$. And $\mathrm{Tor}_1(\underline{o}_x/f\cdot\underline{o}_x, \underline{o}_{x,Z}) = (0)$ since f is not a 0-divisor in $\underline{o}_{x,Z}$ (since f is a unit at all associated primes of $\underline{o}_{x,Z}$). This shows that $\mathcal{I}_H$ is a sheaf of ideals, and we can use induction.

Now, by $(*)_m$,

$$\begin{aligned}\chi(\mathcal{I}_H(m+1)) &= \chi(\mathcal{I}(m+1)) - \chi(\mathcal{I}(m)) \\ &= \sum_{i=0}^{n} a_i[\binom{m+1}{i} - \binom{m}{i}] \\ &= \sum_{i=0}^{n-1} a_{i+1}\binom{m}{i} .\end{aligned}$$

Therefore we can assume that $\mathcal{I}_H$ is $G(a_1, a_2, \ldots, a_n)$-regular, for a suitable polynomial G depending only on n. Put $m_1 = G(a_1, \ldots, a_n)$. Then we get, by $(*)_m$:

(i) $$0 \to H^0(\mathcal{I}(m)) \to H^0(\mathcal{I}(m+1)) \xrightarrow{\rho_{m+1}} H^0(\mathcal{I}_H(m+1)) \to H^1(\mathcal{I}(m)) \to H^1(\mathcal{I}(m+1)) \to 0$$

for $m \geq m_1-2$. And for any $i \geq 2$, we get:

(ii) $$0 \to H^i(\mathcal{I}(m)) \to H^i(\mathcal{I}(m+1)) \to 0$$

for $m \geq m_1-i$.

Now since $H^i(\mathcal{I}(m)) = (0)$, for $i \geq 1$ and $m >> 0$, this last sequence (ii) tells us that $H^i(\mathcal{I}(m)) = (0)$ as soon as $i \geq 2$ and $m \geq m_1-i$. This means that as far as $H^2, H^3, \ldots, H^n$ are concerned, $\mathcal{I}$ is also m_1-regular. On the other hand, sequence (i) tells us:

(#) If $m \geq m_1-2$, then either ρ_{m+1} is surjective or
$$\dim H^1(\mathcal{I}(m+1)) < \dim H^1(\mathcal{I}(m)) .$$

But suppose that for $m = m_2$, where $m_2 \geq m_1$, ρ_{m_2} is surjective. By the Proposition we know that

$$H^0(\mathcal{I}_H(m_2)) \otimes H^0(\underline{o}_H(1)) \to H^0(\mathcal{I}_H(m_2+1))$$

is surjective. Therefore it follows that the image of $H^0(\mathcal{I}(m_2)) \otimes H^0(\underline{o}_{\mathbf{P}_n}(1))$ in $H^0(\mathcal{I}(m_2+1))$ is mapped surjectively onto $H^0(\mathcal{I}_H(m_2+1))$. Hence, *a fortiori*, ρ_{m_2+1} is surjective. In other words, looking at all $m \geq m_1$, once ρ_m is surjective, it is surjective for all larger m. Hence:

(#') If $m \geq m_1 - 1$, $\dim H^1(\mathcal{I}(m))$ is *strictly* decreasing, as a function of m, until it reaches 0.

Therefore clearly:

$$\mathcal{I} \text{ is } [m_1 + \dim H^1(\mathcal{I}(m_1-1))]\text{-regular} .$$

Up to this point, we have not used the fact that $\mathcal{I}$ is a sheaf of ideals. But now we compute:

$$\begin{aligned}\dim H^1(\mathcal{J}(m_1-1)) &= \dim H^0(\mathcal{J}(m_1-1)) - \chi(\mathcal{J}(m_1-1)) \\ &\leq \dim H^0(\underline{o}_{\mathbf{P}_n}(m_1-1)) - \chi(\mathcal{J}(m_1-1)) \\ &= H(a_0, a_1, \ldots, a_n; m_1)\end{aligned}$$

where H is a polynomial in the a's and in m_1. In short, $\mathcal{J}$ is

$$G(a_1, \ldots, a_n) + H(a_0, \ldots, a_n; G(a_1, \ldots, a_n))$$

regular.

QED

A few remarks: First of all, the theorem is false unless $\mathcal{J}$ is assumed to be a sheaf of ideals. Thus, take $n = 1$, and let

$$\mathcal{F}_k = \underline{o}_{\mathbf{P}_1}(+k) \oplus \underline{o}_{\mathbf{P}_1}(-k) \quad .$$

Then $\chi(\mathcal{F}_k(m)) = 2(m+1)$, which is independent of k: but the least m such that $\mathcal{F}_k$ is m-regular is $m = |k| - 1$.

Second, suppose we are concerned with the geometry on a fixed projective algebraic scheme X; then the analogous result is true—

> Fix an immersion $X \subset \mathbf{P}_n$, and say $r = \dim X$; then there is a polynomial $F(x_0, \ldots, x_r)$ such that if $\mathcal{J} \subset \underline{o}_X$ is any sheaf of ideals, and $\chi(\mathcal{J}(m)) = \sum_{i=0}^{r} a_i \binom{m}{i}$, then $\mathcal{J}$ is $F(a_0, \ldots, a_r)$-regular.

To prove this, for a given $\mathcal{J}$, let $\mathcal{J}$ define the closed subscheme $Z \subset X$, hence $Z \subset \mathbf{P}_n$, and let $\mathfrak{J}$ be the sheaf of ideals on $\mathbf{P}_n$ defining Z. Moreover, let K be the sheaf of ideals on $\mathbf{P}_n$ defining X. Then one has the sequence:

$$0 \to K \to \mathfrak{J} \to \mathcal{J} \to 0 \quad .$$

It follows that if $\mathfrak{J}$ is m_0-regular, and $H^i(K(m)) = (0)$, for $i+m = m_0+1$, then $\mathcal{J}$ is m_0-regular as a sheaf on X. But since

$$\chi(\mathfrak{J}(m)) = \chi(\mathcal{J}(m)) + \underbrace{\chi(K(m))}_{\text{independent of } \mathcal{J}}$$

the corollary follows from the theorem. It also follows from the Proposition that $H^0(\mathcal{J}(m_0+k)) \otimes H^0(\underline{o}_X(1)) \to H^0(\mathcal{J}(m_0+k+1))$ is surjective if $k \geq 0$, and that $\mathcal{J}(m)$ is generated by its global sections if $m \geq m_0$.

LECTURE 15

UNIVERSAL FAMILIES OF CURVES

We are now ready to prove that the scheme $C(\xi)$ of Lecture 12 exists. Fix a non-singular projective surface F, and fix an embedding $F \subset \mathbf{P}_n$. As usual, let $\underline{o}(1)$ be the induced very ample invertible sheaf. In Lecture 12, we made the decomposition:

$$\underline{\text{Curves}_F}(S) = \coprod_{\xi \in \text{Num}(F)} \underline{\text{Curves}_F^{\xi}}(S)$$

(for S connected). Actually, for the purposes of this particular proof we will only need a coarser decomposition. In fact, given $D \subset F$, we will only look at the Hilbert polynomial:

$$P(n) = \chi(\underline{o}_F(-D + n)) .$$

In virtue of Proposition 3, Lecture 12, $P(n)$ is determined by the numerical image ξ of D. In fact, $P(n)$ is determined by a) the degree d of D, and b) the arithmetic genus $p_a(D)$. This is seen by

$$(\#) \qquad 0 \to \underline{o}_F(-D + n) \to \underline{o}_F(n) \to \underline{o}_D(n) \to 0 ,$$

hence

$$\begin{aligned} P(n) &= \chi(\underline{o}_F(n)) - \chi(\underline{o}_D(n)) \\ &= \chi(\underline{o}_F(n)) - d \cdot n - 1 + p_a(D). \end{aligned}$$

In any case, we will use the decomposition:

$$\underline{\text{Curves}_F}(S) = \coprod_{P} \underline{\text{Curves}_F^{P}}(S)$$

(for S connected), where $\underline{\text{Curves}_F^P}(S)$ is the set of $D \subset F \times S$ such that $\underline{o}_{F\times S}(-D)$ has Hilbert polynomial P on each fibre. To be precise, if S is not connected, then say $S = \amalg_\alpha S_\alpha$, where S_α is connected, and let

$$\underline{\text{Curves}_F^P}(S) = \prod_{\alpha} \underline{\text{Curves}_F^P}(S_\alpha) .$$

It is very easy to check that this is a subfunctor of $\underline{\text{Curves}_F}$; and if this is represented by a (algebraic) scheme $C(P)$, then $C(P)$ is a disjoint union of open subsets $C(\xi)$ representing the various sub-functors $\underline{\text{Curves}}_F^{\xi}$. Now fix some P.

(I.) By Lecture 14, there is an m_0 depending only on P, such that if $D \subset F$ is any curve giving the Hilbert polynomial P, then $\underline{o}_F(-D)$ is m_0-regular. We may as well also assume that

$$H^1(\underline{o}_F(m_0)) = (0) .$$

Then we conclude:

(a) $H^1(\underline{o}_F(-D + m_0)) = H^2(\underline{o}_F(-D + m_0)) = (0)$,

and $\underline{o}_F(-D + m_0)$ is spanned by its sections.

Using the exact sequence (#) for $n = m_0$, we also conclude:

(b) $H^1(\underline{o}_D(m_0)) = (0)$.

(II.) Now suppose $D \subset F \times S$ is any family of curves giving the Hilbert polynomial P. First of all, we get:

$(b)_S$ $p_*(\underline{o}_D(m_0))$ is locally free, of rank

$$r = \chi(\underline{o}_F(m_0)) - P(m_0) ,$$

(depending only on P), and the formation of p_* commutes with base extensions $T \xrightarrow{g} S$.

This follows from (b), from Corollary 1, 3°, Lecture 7, and from the exact sequence (#).

The useful consequences of (a) will be:

$(a)_S$ $$R^1 p_*(\underline{o}_{F\times S}(-D + m_0)) = (0),$$

and

$$p^* p_*[\underline{o}_{F\times S}(-D + m_0)] \to \underline{o}_{F\times S}(-D + m_0) \quad \text{is surjective} .$$

The first is true by Corollary 1, 3°, Lecture 7; and the second is true because $p_*[\underline{o}_{F\times S}(-D + m_0)]$ maps onto $H^0(\underline{o}_F(-D_s + m_0))$ for all closed points $s \in S$; and $H^0(\underline{o}_F(-D_s + m_0))$ generates $\underline{o}_F(-D_s + m_0) = \underline{o}_{F\times S}(-D + m_0)_{\underline{o}_S} \otimes \mathcal{K}(s)$.

(III.) Again suppose $D \subset F \times S$ is a family of curves. From the sequence (#) for $n = m_0$ and $(a)_S$, we get:

$$0 \to p_*[\underline{o}_{F\times S}(-D + m_0)] \to p_*[\underline{o}_{F\times S}(m_0)] \xrightarrow{\sigma} p_*[\underline{o}_D(m_0)] \to 0$$

$$p_*[\underline{o}_{F\times S}(m_0)] \cong \underline{o}_S \otimes_k H^0(\underline{o}_F(m_0)) .$$

Fixing a basis $e_0, e_1, \ldots, e_N$ of $H^0(\underline{o}_F(m_0))$, we have determined:

a) a locally free sheaf $p_*[\underline{o}_D(m_0)]$ of rank r,

b) N + 1-sections $s_i = \sigma(1 \otimes e_i)$ which span $p_*[\underline{o}_D(m_0)]$.

This is an S-valued point of the Grassmannian $G_{N,r}$! In virtue of $(b)_S$, the formation of $p_*[\underline{o}_D(m_0)]$ is functorial in S, and the whole procedure defines a morphism of functors:

$$\underline{\text{Curves}}^P_F \xrightarrow{\Phi} h_{G_{N,r}} .$$

(IV.) Now suppose we are given an S-valued point, $S \xrightarrow{f} G_{N,r}$ of $G_{N,r}$. Then f defines a locally free sheaf $\mathcal{E}$ of rank r and (N + 1)-sections $s_0, \ldots, s_N$ spanning $\mathcal{E}$. This defines a surjective homomorphism:

$$\underline{o}_S \otimes_k H^0(\underline{o}_F(m_0)) \xrightarrow{\sigma} \mathcal{E} \longrightarrow 0 .$$

Let K be the kernel of σ. Then pulling up via $p\colon F \times S \to S$, we obtain

$$\begin{array}{ccccccc} p^*(K) & \to & p^*[\underline{o}_S \otimes_k H^0(\underline{o}_F(m_0))] & \to & p^*\mathcal{E} & \to & 0 \\ & & \downarrow & & & & \\ & & \underline{o}_{F\times S}(m_0) & & & & \end{array}$$

Define $\mathcal{I}$ to be the image of $p^*(K)(-m_0)$ in $\underline{o}_{F\times S}$: a sheaf of ideals on $F \times S$. This whole procedure defines a morphism of functors:

$$h_{G_{N,r}} \xrightarrow{\Psi} \underline{\text{All}}\ \underline{\text{Subschemes}}_F$$

(V.) What is $\Psi \circ \Phi$? Start with $D \subset F \times S$, and construct f as in (IV.). Then following the procedure of (IV.):

$$\mathcal{E} \cong p_*(\underline{o}_D(m_0))$$

and

$$K \cong p_*(\underline{o}_{F\times S}(-D + m_0)) .$$

But we saw in $(a)_S$ that the subsheaf $\underline{o}_{F\times S}(-D + m_0)$ of $\underline{o}_{F\times S}(m_0)$ was spanned by the sections in this K, i.e., the image of $p^*(K)$ in $\underline{o}_{F\times S}(m_0)$ is exactly $\underline{o}_{F\times S}(-D + m_0)$. Therefore $\mathcal{I}$ is $\underline{o}_{F\times S}(-D)$; i.e.,

$$\Psi \circ \Phi = \left[\begin{array}{l}\text{natural inclusion of} \\ \underline{\text{Curves}} \text{ in } \underline{\text{All}}\ \underline{\text{Subschemes}}\end{array}\right] .$$

(VI.) We can abstract the rest of the argument: given the set-up

$$\begin{array}{ccc} A & \xrightarrow{\Phi} & \\ i \ \cap & & h_G \\ B & \xleftarrow{\Psi} & \end{array}$$

of morphisms of functors (from the category of algebraic schemes /k to the category of sets), assume that:

(#) for all $\alpha \in B(S)$, there is a subscheme $Y \subset S$ such that for all $T \xrightarrow{g} S$,

$$\left(\begin{array}{c} g^*(\alpha) \in B(T) \\ \text{is in the subset} \\ A(T) \end{array}\right) \iff \left(\begin{array}{l} g \text{ factors} \\ \text{through } Y \end{array}\right) .$$

Then there is a subscheme $G_0 \subset G$ such that $A \cong h_{G_0}$, Φ being the inclusion of h_{G_0} in h_G.

(Proof left to the reader.)

(VII.) We must verify (#). In our case, it means

$(\#)_0$ for all closed subschemes $Z \subset F \times S$, there is a subscheme $Y \subset S$ such that for all $T \xrightarrow{g} S$,

$$\left(\begin{array}{l} Z \times_S T \subset F \times T \text{ is a} \\ \text{family of curves} \\ \text{over } T\text{, whose} \\ \text{sheaf of ideals has} \\ \text{Hilbert polynomial } P \end{array}\right) \iff \left(\begin{array}{l} g \text{ factors} \\ \text{through } Y \end{array}\right)$$

But by the key result on flattening stratifications, there is a subscheme $Y \subset S$ such that $Z \times_S T$ is flat over T, with Hilbert polynomial

$$\chi(\underline{o}_{Z\times_S T}(n)) = \chi(\underline{o}_F(n)) - P(n)$$

if and only if g factors through Y. It remains to analyze when $X \times_S T$ is actually a Cartier divisor. This is dealt with by:

<u>Lemma</u>: Let $Z \subset F \times T$ be a closed subscheme, flat over T. Let $t \in T$ be a closed point such that Z_t is a curve on F. Then there is an open neighborhood U of t in T such that $Z \cap (F \times U)$ is a Cartier divisor on U.

<u>Proof</u>: Since $p\colon F \times T \to T$ is a closed map, it suffices to prove that there is an open neighborhood $\mathcal{U}$ of $F \times \{t\}$ in which Z is a Cartier divisor. Let $x \in F \times T$ be any point such that $p(x) = t$. Let $\mathcal{I}_x \subset \underline{o}_x$ be the ideal defining Z at x, and let $m_t \subset \underline{o}_t$ be the maximal ideal. Since $\underline{o}_x/m_t \cdot \underline{o}_x$ is the local ring of x on $F \times \{t\}$, and since Z_t is a Cartier divisor,

$$\mathcal{I}_x + m_t \cdot \underline{o}_x = (f) + m_t \cdot \underline{o}_x$$

for some $f \in \underline{o}_x$. Choosing f suitably, we may assume that $f \in \mathcal{I}_x$. Then look at the exact sequence:

$$0 \to \mathcal{I}_x/(f) \to \underline{o}_x/(f) \to \underline{o}_x/\mathcal{I}_x \to 0 .$$

Since Z is flat over T, we get:

$$\mathrm{Tor}_1^{\underline{o}_t}(\underline{o}_x/\mathcal{I}_x, \mathcal{K}(t)) \to \mathcal{I}_x/(f) \otimes \mathcal{K}(t) \to \underline{o}_x/(f) + m_t\cdot\underline{o}_x \to \underline{o}_x/\mathcal{I}_x + m_t\cdot\underline{o}_x) \to 0 .$$

$$\overset{\displaystyle \|}{(0)}$$

Therefore, $[\mathcal{I}_x/(f)] \otimes \mathcal{K}(t) = (0)$, hence by Nakayama's lemma, $\mathcal{I}_x/(f) = (0)$. This proves that $\mathcal{I}_x = (f)$, hence Z is a Cartier divisor at x, hence in a neighborhood of x.

QED

(VIII.) This proves the first construction theorem: that a universal family of curves exists. One point, however, does not follow from our discussion. We do know that the parameter scheme for this universal family is a subscheme Y of $G_{N,r}$. However, it is even a closed subscheme, hence Y is even a projective scheme.

Proof: Let $\overline{Y}$ be the closure of Y as a subset of $G_{N,r}$. Assume $Y \subsetneq \overline{Y}$. Then pick a closed point $y \in \overline{Y} - Y$. Fix

i) $U = \mathrm{Spec}\ (R)$, an affine neighborhood of $y \in \overline{Y}$,

ii) the maximal ideal $m \subset R$ defining y,

iii) an ideal $\mathfrak{A} \subset m$ defining the closed subset $(\overline{Y} - Y) \cap U$.

Then it is easy to check that there is some prime ideal $\wp$ such that:

$$\wp \subset m, \quad \wp \not\supset \mathfrak{A}, \quad \dim[R_\wp/\wp \cdot R_\wp] = 1 \quad .$$

Let S be the integral closure of the domain $R/\wp$ in its quotient field, and let $C = \mathrm{Spec}\ (S)$. Then C is a 1-dimensional non-singular variety, and the given homomorphism from R to S induces a morphism

$$C \xrightarrow{f} U \subset \overline{Y} .$$

If $\overline{\wp} \subset S$ is a prime ideal lying over $m \cdot (R/\wp)$, then $\overline{\wp}$ defines a closed point $z \in C$ such that $y = f(z)$. Let $C_0 = f^{-1}(Y)$, and let f_0 be the restriction of f to C_0. Then f_0 is a C_0-valued point of Y which is not the restriction of a C-valued point of Y, i.e., because in the closure of the graph of f_0, $f(x) \notin Y$.

We shall show that this is absurd. But h_Y is isomorphic to $\underline{\mathrm{Curves}}_F^P$. Therefore f_0 defines a family of curves

$$D_0 \subset F \times C_0$$

(giving the polynomial P) which is not the restriction of a family of curves over C. But since C and F are non-singular, $F \times C$ is non-singular, and divisors on $F \times C$ are the same as Weil divisors. In particular, let D_0, as a Weil divisor, be written out as:

$$D_0 = \sum n_i Z_{i,0}$$

where $Z_{i,0}$ is a closed subset of $F \times C_0$ of codimension 1. Let Z_i be the closure of $Z_{i,0}$ in $F \times C$. Let $D = \Sigma n_i Z_i$. Then D is certainly an effective divisor on $F \times C$. Moreover, it is a relative divisor over C because its support does not contain any of the fibres $F \times \{z\}$, $z \in C$. Therefore D is a family of curves over C extending D_0. Finally, since C is connected, the Hilbert polynomial of $\underline{o}_F(-D)$ is constant, hence equal to P. This contradiction proves the theorem.

LECTURE 16

THE METHOD OF CHOW SCHEMES

Since the existence of these universal families has such pivotal importance in the proof of the main existence theorems, it seems reasonable to sketch the only other known approach to their construction—that of Chow and van der Waerden. Again let $F \subset \mathbf{P}_n$ be given. In this approach we only fix the degree d of a curve $D \subset F$, not the polynomial P as above, i.e., we decompose:

$$\underline{\text{Curves}_F}(S) = \coprod_{d \geq 0} \underline{\text{Curves}_F^d}(S)$$

(for S connected), where $\underline{\text{Curves}_F^d}(S)$ stands for the set of $D \subset F \times S$ such that the induced curves D_s on the fibres all have degree d.

Say X is a projective scheme: then we can define a functor:

$$\underline{\text{Div}_X}(S) = \left\{ \mathfrak{D} \subset X \times S \;\middle|\; \begin{array}{l}\mathfrak{D} \text{ a relative effective Cartier} \\ \text{divisor over } S\end{array} \right\}$$

generalizing $\underline{\text{Curves}_F}$. In some cases where $\dim (X) > 2$, this may be easier to study than $\underline{\text{Curves}_F}$ for some surfaces F. For example, if X is a Grassmannian G, the methods of Lecture 13 enable one to prove that

$$\underline{\text{Div}_G} \cong h_D$$

where D is a disjoint union of projective spaces. In fact, $\underline{\text{Div}_G}$ is broken up into $\underline{\text{Div}_G^k}$, one for each integer $k \geq 0$, and $\underline{\text{Div}_G^k}$ is just a linear system.

The method of Chow is to construct a morphism of functors:

$$\underline{\text{Curves}_F^d} \xrightarrow{\Phi} \underline{\text{Div}_G^d}$$

for the Grassmannian $G = G_{n,n-1}$. To do this, we first construct a subscheme

$$Z \subset \mathbf{P}_n \times G_{n,n-1} .$$

Heuristically, every closed point of $G_{n,n-1}$ corresponds to a linear subspace $L \subset \mathbf{P}_n$ of dimension $n-2$. Putting these together, they form Z. To be precise, recall from Lecture 5 that $G_{n,n-1} = \text{Proj}\,(R)$, where R is a graded ring generated by elements

$$p_{i_1,i_2,\ldots,i_{n-1}}, \quad 0 \le i_1 < i_2 < \ldots < i_{n-1} \le n .$$

If $j < k$ are the two integers omitted in the sequences of i's, we can simplify notation by putting

$$q_{j,k} = p_{i_1,i_2,\ldots,i_{n-1}} \quad .$$

Then Z is defined as the scheme of zeroes of the sections

$$s_k = \left\{ \sum_{j=0}^{k-1} (-1)^j X_j \otimes q_{j,k} - \sum_{j=k+1}^{n} (-1)^j X_j \otimes q_{k,j} \right\}$$

of $\underline{p}_1^*(\underline{o}(1)) \otimes p_2^*(\underline{o}(1))$, for $0 \le k \le n$. Then, in fact, Z is a bundle of $\mathbf{P}_{n-2}$'s over $G_{n,n-1}$, and also a bundle of $G_{n-1,n-2}$'s over $\mathbf{P}_n$. Classically, Z is called the incidence correspondence, and Z itself is a flag manifold.

Now form the fibre product $\mathfrak{X} = F \times_{\mathbf{P}_n} Z$: :

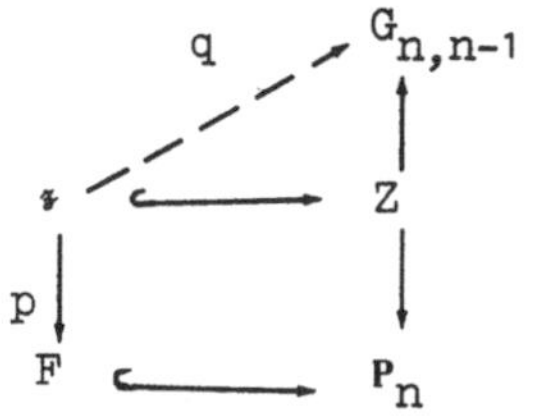

(i) p is a flat morphism; in fact, $\mathfrak{X}$ is a bundle of $G_{n-1,n-2}$'s over F in the sense that F admits an open covering $\{U_i\}$ such that $p^{-1}(U_i) \cong U_i \times G_{n-1,n-2}$. In particular,

$$\begin{aligned} \dim \mathfrak{X} &= \dim F + \dim G_{n-1,n-2} \\ &= 2 + 2(n-2) \\ &= 2(n-1) \\ &= \dim G_{n,n-1} \quad , \end{aligned}$$

and $\mathfrak{X}$ is non-singular.

(ii) Moreover q is a surjective morphism of two non-singular varities of the same dimension. This implies that there is an open subset $U \subset G_{n,n-1}$ containing all points of co-dimension 1 over which q is finite and flat.

(iii) More generally, you can make any base extension to obtain a situation:

$$\begin{array}{ccc} & \mathfrak{z} \times S & \\ {}^{p}\swarrow & & \searrow^{q} \\ F \times S & & G_{n,n-1} \times S \end{array} \quad .$$

One still has:

$$\left\{\begin{array}{l} p \text{ flat} \\ q \text{ of finite Tor-dimension} \\ \text{there exists open subset } U \subset G_{n,n-1} \times S \\ \quad \text{containing all points of depth 1,} \\ \quad \text{over which } q \text{ is finite and flat.} \end{array}\right.$$

Therefore, if $D \subset F \times S$ is a family of curves over S, we can form:

$$\Phi(D) = q_* p^*(D)$$

according to 1° and 3°, Lecture 10.

The rest of the work consists in showing, as in Lecture 15, that Φ is injective, and that if $\underline{\mathrm{Div}}_G^d \cong h_{\mathbf{P}_N}$, then there exists a closed subscheme $Y \subset \mathbf{P}_N$ such that an S-valued point of $\underline{\mathrm{Div}}_G^d$ is in the image of Φ if and only if the corresponding point of $\mathbf{P}_N$ is a point of Y. Then it follows that $\underline{\mathrm{Curves}}_F^d \cong h_Y$. Even the method is similar to that of Lecture 15: one constructs an "inverse" morphism:

$$\Psi: \ \underline{\mathrm{Div}}_G^d \to \underline{\mathrm{All}}\ \underline{\mathrm{subschemes}}_F$$

and then applies the same categorical argument as in part (VI.), Lecture 15. In some sense, the deepest part of the argument is the same—the invoking of the existence of flattening stratifications to verify the hypothesis in the categorical argument.

An interesting corollary of this approach is the stronger finiteness theorem that it yields: for any given degree d, there are only a finite number of elements $\xi \in \mathrm{Num}(F)$ such that:

a) $\deg \xi = d$,

b) ξ is represented by a curve.

The essential facts behind this finiteness are quite interesting and useful. What we want to do is to prove completely a closely related result which seems to contain the key point, and which we will use subsequently.

THEOREM: Let $D \subset F$ be a curve of degree d. Then $\underline{o}_F(-D + d)$ is spanned by its sections.

Proof: We are given an embedding $F \subset \mathbf{P}_n$ inducing the sheaf $\underline{o}(1)$. Suppose $L \subset \mathbf{P}_n$ is a linear subspace of dimension $n-3$. Then recall that there is a "projection"

$$\pi: \ (\mathbf{P}_n - L) \to \mathbf{P}_2 \quad .$$

[In our approach, we can define π as a $(\mathbf{P}_n - L)$-valued point of $\mathbf{P}_2$. Namely, let $L = H_1 \cap H_2 \cap H_3$, where H_i is the hyperplane defined by $h_i \in H^0(\mathbf{P}_n, \underline{o}(1))$. Then the three sections h_1, h_2 and h_3 of $\underline{o}(1)$ have no common zeroes in $\mathbf{P}_n - L$, and they define the point π.]

In particular, if $F \cap L = \emptyset$, then π restricts to a morphism

$$\pi' : F \to \mathbf{P}_2 \quad .$$

I claim that π' is finite and flat.

a) π is affine: $\mathbf{P}_2$ is covered by affine open sets $\mathbf{P}_2 - \ell_i$ (ℓ_1, ℓ_2, ℓ_3 the three fundamental lines) and $\pi^{-1}(\mathbf{P}_2 - \ell_i) = \mathbf{P}_n - H_i$ is affine.

b) Therefore π' is affine because π' is the restriction of π to a closed subscheme.

c) Let i denote the inclusion of F in $\mathbf{P}_n$. Then π' factors:

$$\begin{array}{ccc} F & \xrightarrow{(i,\ \pi')} & \mathbf{P}_n \times \mathbf{P}_2 \\ & \searrow^{\pi'} & \downarrow p_2 \\ & & \mathbf{P}_2 \end{array}$$

Since (i, π') is an isomorphism of F with a closed subscheme of $\mathbf{P}_n \times \mathbf{P}_2$, the direct image sheaf $\pi'_*(\underline{o}_F)$ is just the same as $p_{2,*}(\underline{o}_F)$ (where $\underline{o}_F$ is identified with the structure sheaf of the image of F in $\mathbf{P}_n \times \mathbf{P}_2$). This is coherent (cf. 2°, Lecture 7). Therefore π' is finite.

d) The fact that π' is flat follows from the general result:

Lemma: Let A be a regular local ring of dimension n, and let B be an A-aglebra, finitely generated as A-module. If all localizations of B with respect to maximal ideals are Cohen-Macaulay rings of dimension n, then B is a free A-module.

(Cf. NAGATA, Local Rings, (25.16), and EGA 4, § 15.4.)

Now suppose that $D \subset F$ is a curve of degree d and π' is such a morphism. Then $\pi'_*(D)$ is defined by Norms, as in 2°, Lecture 10. This is a plane curve, and I claim that its degree is d too.

Computation of deg $\pi'_*(D)$:

Start with a line $\ell \subset \mathbf{P}_2$ which doesn't contain any of the generic points of the set $\pi'_*(D)$: then $\ell \cap \operatorname{Supp} \pi'_*(D)$ is 0-dimensional, and

$$\deg \pi'_*(D) = (\ell \cdot \pi'_*(D)) \quad .$$

Let $\{x_1, \ldots, x_n\} = \ell \cap \operatorname{Supp} \pi'_*(D)$. At each point x_i, let $\underline{o}_i = \underline{o}_{x_i}$,

$f_i \in \underline{o}_i$ a local equation of ℓ, $R_i = [\pi^!_*(\underline{o}_F)]_{x_i}$, $g_i \in R_i$ a local equation of D in a neighborhood of the set $\pi'^{-1}(x_i)$. Then R_i is a finitely generated free $\underline{o}_i$-module, and $\mathrm{Nm}(g_i)$ is a local equation of $\pi^!_*(D)$. Moreover,

$$(\ell \cdot \pi^!_*(D)) = \sum_{i=1}^{n} \dim_k \underline{o}_i/(f_i, \mathrm{Nm}\, g_i) \quad .$$

By an elementary result on determinants*, we get

$$\dim_k \underline{o}_i/(f_i, \mathrm{Nm}\, g_i) = \dim_k R_i/(f_i, g_i)$$

and, by definition:

$$\begin{aligned}\sum_{i=1}^{n} \dim_k R_i/(f_i, g_i) &= (\pi'^*(\ell) \cdot D)\\ &= (\underline{o}(1), \underline{o}_F(D))\\ &= \deg D\\ &= d \quad .\end{aligned}$$

We now come to the main point:

$$\pi'^*(\pi^!_*(D)) = D + D'$$

where D' is effective, by statement (*), 2°, Lecture 10! And, in fact, since the divisor class of $\pi^!_*(D)$ is $\underline{o}(d)$, the divisor class of $\pi'^*(\pi^!_*(D))$ is also $\underline{o}(d)$, hence the divisor class of D' is $\underline{o}_F(-D + d)$. The theorem, therefore, will be proven if we can show the following

(*) For all closed points $x \in F$, there is a linear space L of dimension n-3 such that $L \cap F = \emptyset$, and such that the divisor D', constructed as above, does not pass through x.

In other words, we require:

$$\pi'^*(\pi^!_*(D))_x = D_x \quad .$$

First of all, let's analyze what we need to get this out: let $\underline{o}$ be the local ring of $\mathbf{P}_2$ at $\pi'(x)$, and let R be the stalk of $\pi^!_*(\underline{o}_F)$ at $\pi'(x)$. Let $g \in R$ be a local equation of D at all points $\pi'^{-1}(\pi'(x))$, and let $m \subset R$ be the maximal ideal such that R_m is the local ring of F at x.

---

* Let A be a 1-dimensional local ring, M a free A-module of finite type, $T\colon M \to M$ an A-linear injective homomorphism. Then:

$$\text{length}\,(M/T(M)) = \text{length}\,(A/(\det T)) \; .$$

Passing to the completions, we find

$$\hat{R} = R \otimes_{\underline{o}} \hat{\underline{o}} \cong (\widehat{R_m}) \oplus \sum_{\substack{\text{other maximal} \\ \text{ideals } m' \subset R}} (\widehat{R_{m'}}) .$$

The image of Nm (g) in $\hat{\underline{o}}$ is then the product of the Norms of g from <u>each</u> component $(\widehat{R_{m'}})$ to $\hat{\underline{o}}$. But we want g and Nm (g) to differ by a unit. Therefore, first we need:

a) g is a unit at all other localizations $R_{m'}$ of R; i.e., Supp (D) does not contain any points $x' \neq x$ such that $\pi'(x') = \pi'(x)$.

If this holds, the image Nm (g) in $\hat{\underline{o}}$ is just the Norm from $(\widehat{R_m})$ to $\hat{\underline{o}}$ Therefore, secondly we can use

b) $\hat{\underline{o}} \cong \widehat{R_m}$; i.e., R_m is <u>unramified</u> over $\underline{o}$, or equivalently the map from the Zariski tangent space to F at x to the Zariski tangent space to $\mathbf{P}_2$ at $\pi'(x)$ is an isomorphism.

If this holds, Nm (g) and g differ only by a unit in $\widehat{R_m}$; therefore they differ only by a unit in R_m.

What are the corresponding geometric conditions on L ? Clearly a) becomes:

a') If $\tilde{L}$ is the linear space of dimension n-2 spanned by L and x, then x is the only intersection of $\tilde{L}$ and Supp (D).

On the other hand, look at the Zariski tangent space T to $\mathbf{P}_n$ at x; this contains the tangent space $T_{\tilde{L}}$ to $\tilde{L}$, of dimension n-2, and the tangent space T_F to F, of dimension 2. Moreover, the full projection π induces an isomorphism of $T/T_{\tilde{L}}$ with the tangent space to $\mathbf{P}_2$ at $\pi(x)$. Therefore b) becomes:

b') The tangent spaces $T_{\tilde{L}}$ and T_F to $\tilde{L}$ and F intersect transversely at x.

The rest is easy: let M be the 2-dimensional linear space through x with tangent space T_F at x. First choose $h \in H^0(\mathbf{P}_n, \underline{o}(1))$ such that

$$\begin{cases} h(x) = 0 \\ h(y) \neq 0, & \text{for } y \text{ the generic point of } M \text{ or for} \\ & y \text{ a generic point of Supp (D).} \end{cases}$$

Let H be the corresponding hyperplane. Second, choose $h' \in H^0(\mathbf{P}_n, \underline{o}(1))$ such that

$$\begin{cases} h'(x) = 0 \\ h'(y) \neq 0, & \text{for } y \text{ the generic point of } M \cap H \\ & \text{or for } y \text{ a generic point of } F \cap H \\ & \text{or for } y \in \{\mathrm{Supp}\,(D) \cap H\} - \{x\}. \end{cases}$$

Let H' be the corresponding hyperplane. Let $\tilde{L} = H \cap H'$. Then $\tilde{L}$ satisfied a') and b') and $\tilde{L} \cap F$ is 0-dimensional. Let L be a linear subspace of $\tilde{L}$ of dimension n-3 not containing any of the finite set of points $\tilde{L} \cap F$.

QED

The corollary of the theorem which can be used to bound $\chi(\underline{o}_F(D))$ in terms of $\deg(D)$ is this:

If D is a curve on F, then
$$(D \cdot D) \geq -A \cdot \deg(D)^2$$
where $A = (\underline{o}(1) \cdot \underline{o}(1)) - 2$.

We omit the proof since we have no other applications for this fact.

LECTURE 17

GOOD CURVES

In this lecture, we want to give a partial answer to the third question posed in Lecture I: What is a good curve on our surface F ? More precisely, we don't want to distinguish between linearly equivalent curves, so the question becomes—what is a good divisor class on F ? The point is this: Given an arbitrary invertible sheaf L, for very large n the sheaf $L(n)$ should have every "good" property one can ask for. Also look at the analogous question on a curve C (C reduced and irreducible for example). Then an invertible sheaf L on C is "good" if its degree is large enough.

1° Let's be precise: fix once and for all an embedding $F \subset \mathbf{P}_n$ and let $\underline{o}(1)$ be the induced invertible sheaf. Then the set of divisor classes $\underline{\text{Pic}}(F)$ has a fixed automorphism: $L \mapsto L(1)$. The following are various good properties for L:

(I.) L is 0-regular: $H^i(L(n)) = (0)$ if $i + n = 0$, $i > 0$, [hence $H^i(L(n)) = (0)$ if $i + n \geq 0$, $i > 0$].

(II.) L is spanned by its sections; equivalently, for every closed point $x \in F$, there is a curve $D \subset F$ such that

$$\begin{cases} \underline{o}_F(D) \cong L \\ x \notin \text{Supp}(D) . \end{cases}$$

(III.) L is very ample.

(IV.) There is a curve $D \subset F$ with no multiple components such that

$$\underline{o}_F(D) \cong L .$$

What is the relationship between these various properties? Note first of all, that if L has any of these properties, then $L(n)$ has the same property for all $n \geq 0$.

Proof: This is clear for (I.) and (II.). For (III.) we need:

LEMMA A: Let L and M be two invertible sheaves on F. Assume L is spanned by its sections and M is very ample. Then $L \otimes M$ is very ample.

Proof of Lemma: Since L is spanned by its sections, there is a morphism $\varphi\colon F \to \mathbf{P}_{m_1}$ such that $L \cong \varphi^*(\underline{o}(1))$; since M is very ample, there is a closed immersion $\psi\colon F \to \mathbf{P}_{m_2}$ such that $M \cong \psi^*(\underline{o}(1))$. Together these define a closed immersion

$$(\varphi, \psi)\colon F \to \mathbf{P}_{m_1} \times \mathbf{P}_{m_2} .$$

On the other hand, one has the canonical Segre immersion

$$i\colon \mathbf{P}_{m_1} \times \mathbf{P}_{m_2} \subset \mathbf{P}_{m_1m_2+m_1+m_2} .$$

This is defined by the requirements:

$$\begin{cases} i^*(\underline{o}(1)) = p_1^*(\underline{o}(1)) \otimes p_2^*(\underline{o}(1)), \\ i^*(X_j), \text{ for } 0 \leq j \leq m_1m_2 + m_1 + m_2 , \\ \text{are the sections } p_1^*(X_k) \otimes p_2^*(X_\ell) \text{ for} \\ 0 \leq k \leq m_1, \quad 0 \leq \ell \leq m_2, \text{ in some order.} \end{cases}$$

(Exercise: check that this is a closed immersion.) Therefore, $i \circ (\varphi, \psi)$ is a closed immersion of F in $\mathbf{P}_{m_1m_2+m_1+m_2}$ and

$$\begin{aligned} [i \circ (\varphi, \psi)]^*(\underline{o}(1)) &= (\varphi, \psi)^*(p_1^*(\underline{o}(1)) \otimes p_2^*(\underline{o}(1))) \\ &= \varphi^*(\underline{o}(1)) \otimes \psi^*(\underline{o}(1)) \\ &= M \otimes L . \end{aligned}$$

QED

On the other hand, suppose L has property (IV.). We shall use without proof an elementary form of Bertini's Theorem:

LEMMA B: Let L be a very ample invertible sheaf on F. Then there is a non-singular irreducible curve $D \subset F$ such that $L \cong \underline{o}_F(D)$.

Now, if L has Property (IV.), $L \cong \underline{o}_F(D)$, and $D = \Sigma_{i=1}^n D_i$, where the D_i are distinct irreducible curves. Suppose the divisor classes of the curves $D_1, \ldots, D_{n_0}$ are multiples (necessarily positive) of $\underline{o}(1)$, whereas the divisor classes of the other components $D_{n_0+1}, \ldots, D_n$ are not. Say

$$\underline{o}_F\left(\sum_{i=1}^{n_0} D_i\right) \cong \underline{o}(r) .$$

By Lemma A, $\underline{o}(r+1)$ is very ample; by Lemma B, $\underline{o}(r+1) \cong \underline{o}_F(E)$ for some irreducible curve E. Then

$$L(1) \cong \underline{o}_F\left(E + \sum_{i=n_0+1}^{n_0} D_i\right)$$

and all the curves $E, D_{n_0+1}, \ldots, D_n$ are distinct since the sheaves $\underline{o}_F(E)$ and $\underline{o}_F(D_i)$ are not isomorphic for $i > n_0$. This proves that $L(1)$ has property (IV.).

Therefore, all the "good" properties (I.)–(IV.) are stable, in the sense that replacing L by $L(1)$ never destroys them. Our main result is that they are nearly equivalent:

THEOREM 1: There is an integer k depending only on F, and its embedding $F \subset P_n$, such that if an invertible sheaf L is good in any of the senses (I.)–(IV.), then $L(k)$ is good in all four senses.

Proof: We prove this in a chain:

First step: If L is good in sense (I.), then by the Proposition of Lecture 14, L is good in sense (II.). If L is good in sense (II.), then by Lemma A, $L(1)$ is good in sense (III.). If L is good in sense (III.), then by Lemma B, L is good in sense (IV.).

Second step: It remains to get back from (IV.) to (I.). Suppose $L = \underline{o}_F(D)$, where D has no multiple components. Tensor the exact sequence:

$$0 \to \underline{o}_F(-D) \to \underline{o}_F \to \underline{o}_D \to 0$$

with $L(n)$ to obtain:

(*) $$0 \to \underline{o}_F(n) \to L(n) \to L_D(n) \to 0$$

where $L_D = L \otimes \underline{o}_D$. Let n_0 be an integer such that

$$H^i(\underline{o}_F(n)) = (0), \quad n \geq n_0, \quad i > 0 .$$

Then if $n \geq n_0$, it follows from (*) that

i) $$\begin{cases} H^2(L(n)) = (0) \\ H^1(L(n)) \cong H^1(L_D(n)) \end{cases} .$$

We use the Riemann-Roch Theorem on D to attack this last group: let Ω be the canonical sheaf on F (Theorem 3, Lecture 12). Then $\Omega_D = [\Omega \otimes \underline{o}_F(D)] \otimes \underline{o}_D$ and

ii) $$\dim H^1(L_D(n)) = \dim H^0(\Omega_D \otimes L_D^{-1}(-n)) .$$

But iii) $$\Omega_D \otimes L_D^{-1}(-n) = [\Omega \otimes \underline{o}_F(D) \otimes L^{-1} \otimes \underline{o}(-n)] \otimes \underline{o}_D = [\Omega(-n) \otimes \underline{o}_D] .$$

Now there certainly is an integer n_1 such that

$$H^i(\Omega^{-1}(n)) = (0), \quad n \geq n_1, \quad i > 0 \ .$$

By what we proved in the first step, it follows that $\Omega^{-1}(n)$ is very ample if $n \geq n_1+3$. Assume that $n \geq n_1+3$: then the induced invertible sheaf $M = \Omega^{-1}(n) \otimes \underline{o}_D$ on D is very ample on D, i.e., induced from an embedding $i\colon D \hookrightarrow \mathbf{P}_n$. But note:

LEMMA C: Let X be a closed reduced subscheme of $\mathbf{P}_n$ all of whose components have positive dimension. Then $H^0(\underline{o}_X \otimes \underline{o}_{\mathbf{P}_n}(-1)) = (0)$.

Proof of Lemma: Let $X_1,\ldots, X_n$ be the components of X. Since

$$\underline{o}_X \subset \bigoplus_{i=1}^{n} \underline{o}_{X_i} \ ,$$

if $\underline{o}_X(-1)$ had a global section, then for some i, $\underline{o}_{X_i}(-1)$ would have a global section. Since X_i is a variety, the only global sections of $\underline{o}_{X_i}$ are constants. Let H be a hyperplane not containing the generic point of X_i: then

$$\underline{o}_{X_i}(-1) \cong \underline{o}_{X_i} \otimes \underline{o}_{\mathbf{P}_n}(-H) \subset \underline{o}_{X_i} \ .$$

Therefore the constant section of $\underline{o}_{X_i}$ must be a section of $\underline{o}_{X_i}(-1)$, i.e., X_i must be disjoint from H. Then X_i is a closed subscheme of $\mathbf{P}_n - H$, i.e., X_i is finite over k, hence 0-dimensional.

QED

By the lemma,

$$H^0(\Omega(-n) \otimes \underline{o}_D) = (0)$$

if $n \geq n_1+3$. Putting i), ii) and iii) together, it follows that $L(n)$ is 0-regular if

$$n \geq \max[n_0 + 2, \ n_1 + 4] \ .$$

QED

2° This clarifies in a general way the meaning of a "good" curve. The next question is whether there are numerical criteria that imply that an invertible sheaf L is represented by a good curve. In this direction, one has:

Vanishing Lemma D: There is a constant c_1, depending only on F and the very ample sheaf $\underline{o}(1)$ such that, for all invertible sheaves L:

$$\deg(L) \geq c_1 \Longrightarrow H^2(F, L) = (0) \ .$$

Proof: Let $\underline{o}(1) \cong \underline{o}_F(H)$ where H is a non-singular irreducible curve on F. For all k, one has the usual sequence:

$$(\#) \qquad 0 \to L(k-1) \to L(k) \to (L \otimes \underline{o}_H)(k) \to 0 .$$

If $\deg(L) = \deg_H(L \otimes \underline{o}_H) > 2p_a(H) - 2$, then by the vanishing theorem of Lecture 11, $H^1(L \otimes \underline{o}_H) = (0)$. A fortiori, $H^1(L \otimes \underline{o}_H(k)) = (0)$ too, for every integer $k > 0$. Therefore, one concludes:

$$H^2(L) \xrightarrow{\sim} H^2(L(1)) \xrightarrow{\sim} H^2(L(2)) \xrightarrow{\sim} \ldots\ldots .$$

Since for very large n, $H^2(L(n)) = (0)$, it follows that $H^2(L) = (0)$.

QED

COROLLARY 1: With c_1 as above, if an invertible sheaf L satisfies $\deg(L) \geq c_1$, $\chi(L) > 0$, then there is a curve $D \subset F$ such that $L \cong \underline{o}_F(D)$.

COROLLARY 2: With c_1 as above, and $h = \deg \underline{o}(1)$, if an invertible sheaf L satisfies $\deg(L) \geq c_1 + h$, and $H^1(F, L) = (0)$, then $L(2)$ is "good" in all senses.

Proof: By Corollary 1, $H^2(F, L(-1)) = (0)$, hence $L(1)$ is 0-regular, hence $L(2)$ is very ample.

QED

With this, together with the result at the end of Lecture 16, one can prove:

THEOREM 2: There is a constant c_2, and a positive ϵ depending only on F and $\underline{o}(1)$ with the following property: If an invertible sheaf L on F satisfies:

a) $\deg(L) \geq c_2$

b) $\chi(L) \geq (1 - \epsilon)/(2(\underline{o}(1) \cdot \underline{o}(1))) \cdot (\mathrm{Deg}\ L)^2$

then L is 0-regular and very ample.

Proof: Let c_1 be given by Lemma D, let c_3 be the constant of Theorem 1, and let

$$h = (\underline{o}(1) \cdot \underline{o}(1))$$

$$p = \chi(\underline{o}_F) - \chi(\underline{o}(-1))$$

$$= \chi(\underline{o}_H) .$$

Let η be a positive number such that

$$\eta \leq \frac{1}{2h[1 + h(h-2)]} ,$$

let

$$\varepsilon = \frac{(h\eta)^2}{2} ,$$

let

$$c_2 = \max\left\{ \frac{c_1}{\eta h} ; \quad 2h(c_3+1+h(h-2)) ; \quad \frac{1-\eta h}{\varepsilon}(3h+2p); \; -p \right\} .$$

Finally, put

$$k = \left[\frac{\deg L}{h}(1 - \eta \cdot h) \right] .$$

Step I. $\deg L(-k) \geq c_1$.

Proof:

$$\begin{aligned} \deg L(-k) &= \deg L - k \cdot \deg \underline{o}(1) \\ &= \deg L - k \cdot h \\ &\geq \deg L - \left(\frac{\deg L}{h}\right)(1 - \eta \cdot h) \cdot h \\ &\geq c_2 \cdot \eta \cdot h \\ &\geq c_1 \quad . \end{aligned}$$

Step II. $\chi(L(-k)) > 0$.

Proof: Let H be a curve such that $\underline{o}(1) \cong \underline{o}_F(H)$. Use the exact sequence (#) in Lemma D and the Riemann-Roch Theorem on H to obtain the formulae:

$$\begin{aligned} \chi(L(-k)) &= \chi(L) - \sum_{i=0}^{k-1} \chi(L \otimes \underline{o}_H(-i)) \\ &= \chi(L) - k \cdot \chi(\underline{o}_H) - \sum_{i=0}^{k-1} \deg_H\{L \otimes \underline{o}_H(-i)\} \\ &= \chi(L) - k \cdot p - k \deg L + \frac{k(k-1)}{2} \cdot h \quad . \end{aligned}$$

Substituting all our estimates, you get:

$$\begin{aligned} \chi(L(-k)) &\geq \frac{(\deg L)^2}{4} h\eta^2 - (1-\eta h)\frac{\deg L}{2h}(2p + 3h) + h \\ &> \varepsilon \frac{\deg L}{2h}\left[c_2 - \frac{1-\eta h}{\varepsilon}(2p+3h)\right] \geq 0 \quad . \end{aligned}$$

Step III. It follows from Corollary 1 that $L(-k) \cong \underline{o}_F(D)$ for some curve $D \subset F$. Now use the results of Lecture 16: let $d = \deg(D)$. Then $\underline{o}_F(-D + d)$ is spanned by its sections. In particular, there is a curve $E \subset F$ such that $\underline{o}_F(-D + d) \cong \underline{o}_F(E)$. Also,

$$\deg E = -\deg D + d \cdot h = d(h-1) \quad .$$

Again, by that theorem $\underline{o}_F(-E + d(h-1))$ is spanned by its sections. Now

$$\begin{aligned}\underline{o}_F(-E + d(h-1)) &\cong \underline{o}_F(-D + d)^{-1} \otimes \underline{o}_F(d(h-1)) \\ &\cong \underline{o}_F(D) \otimes \underline{o}_F(d(h-2)) \\ &\cong L(d(h-2) - k) .\end{aligned}$$

Therefore, by Theorem 1, $L(d(h-2) - k + c_3)$ is 0-regular and very ample. Therefore the theorem is proven once you deduce:

Step IV: $d(h-2) - k + c_3 \leq 0$.

Proof: Note that $d = \deg D = \deg L - k \cdot h$ so that

$$\begin{aligned} d(h-2) - k + c_3 &= \deg L \cdot (h-2) + c_3 - k(1+h(h-2)) \\ &< \frac{\deg L}{h} \{-1 + \eta h(1+h(h-2))\} + c_3 + 1 + h(h-2) \\ &\leq -\frac{c_2}{2h} + c_3 + 1 + h(h-2) \\ &\leq 0 . \end{aligned}$$

QED

The important thing about this criterion is that, for any invertible sheaf L, the two conditions will be satisfied for $L(n)$ if $n >> 0$. [This is not quite obvious, but it is an exercise.]

COROLLARY: Let $h, \omega \in \mathrm{Num}(F)$ be the images of $\underline{o}(1)$ and of Ω. There are constants c_2 and ε such that if an element $\lambda \in \mathrm{Num}(F)$ satisfies:

i) $(\lambda \cdot h) \geq c_2$

ii) $(\lambda \cdot \lambda - \omega) \geq (1-\varepsilon)\left[\dfrac{(\lambda \cdot h) \cdot (\lambda \cdot h)}{(h \cdot h)}\right]$

then all $L \in \mathrm{Pic}(F)$ representing λ are 0-regular and very ample.

Proof: Use the above Theorem and Proposition 3, Lecture 12 (decreasing ε if necessary).

LECTURE 18

THE INDEX THEOREM

The index theorem for curves on surfaces is a fairly easy Corollary of the theory developed so far. We follow an idea of Grothendieck's (Crelle's Journal, 1958, p. 200).

<u>Proposition</u>: Let L be an invertible sheaf on F such that $(L \cdot L) > 0$. Then

$$[\deg L > 0] \Longleftrightarrow [\text{for some positive } n,\ H^0(F, L^n) \neq (0)] .$$

<u>Proof</u>: If $H^0(F, L^n) \neq 0$, then $L^n \cong \underline{o}_F(D)$ for some curve $D \subset F$. Therefore

$$\begin{aligned} \deg L &= \frac{1}{n} (L^n \cdot \underline{o}(1)) \\ &= \frac{1}{n} (\underline{o}_F(D) \cdot \underline{o}(1)) \\ &= \frac{1}{n} \deg D \\ &> 0 . \end{aligned}$$

Conversely, if $\deg L > 0$, then $H^2(F, L^n) = (0)$ for all sufficiently large n by the vanishing lemma of Lecture 17. Moreover, by Proposition 3, Lecture 12:

$$\begin{aligned} \chi(L^n) &= \frac{1}{2} (L^n \cdot L^n \otimes \Omega^{-1}) + \chi(\underline{o}_F) \\ &= \frac{n^2}{2} (L \cdot L) - \frac{n}{2} (L \cdot \Omega) + \chi(\underline{o}_F) . \end{aligned}$$

This is positive for all sufficiently large n, hence $H^0(F, L^n) \neq (0)$ for all sufficiently large n.

QED

COROLLARY: Let L be an invertible sheaf on F such that $(L \cdot L) > 0$. Then if M_1 and M_2 are two very ample invertible sheaves on F,

$$[(L \cdot M_1) > 0] \Longleftrightarrow [(L \cdot M_2) > 0] .$$

Proof: The point is that the condition $H^0(F, L^n) \neq (0)$ is independent of the given very ample sheaf $\underline{o}(1)$, whereas, by definition, $\deg(L) = (L \cdot \underline{o}(1))$. Therefore the condition "$\deg(L) > 0$" must actually be independent of the choice of very ample sheaf $\underline{o}(1)$.

QED

Index Theorem: Consider the vector space $\text{Num}(F) \otimes \mathbf{Q}$. Let $h \in \text{Num}(F)$ represent the image of $\underline{o}(1)$. Write:

$$\text{Num}(F) \otimes \mathbf{Q} = \{\mathbf{Q} \cdot h\} \oplus \{\mathbf{Q} \cdot h\}^{\perp} .$$

On the second factor, $\{\mathbf{Q} \cdot h\}^{\perp}$, the intersection pairing is negative definite.

Proof: By definition, the pairing on $\text{Num}(F) \otimes \mathbf{Q}$ is non-degenerate. Therefore, it is also non-degenerate on $\{\mathbf{Q} \cdot h\}^{\perp}$. If the theorem were false, there would be an element $k \in \{\mathbf{Q} \cdot h\}^{\perp}$ such that $(k \cdot k) > 0$. Suppose the multiple $a \cdot k$ is represented by an invertible sheaf L. Then $L^n(m)$ represents $m \cdot h + n \cdot a \cdot k$, and

$$\begin{aligned}(L^n(m) \cdot L^{n'}(m')) &= (m \cdot h + n \cdot a \cdot k, m' \cdot h + n' \cdot a \cdot k) \\ &= m \cdot m'(h, h) + n \cdot n' \cdot a^2(k, k) \quad .\end{aligned}$$

In particular, $(L^n(m) \cdot L^n(m)) > 0$ whenever $(n, m) \neq (0, 0)$. Therefore, by the Corollary $(L^n(m) \cdot M)$ is positive for all very ample sheaves M if it is positive for one such M.

Now, $\underline{o}(1)$ is very ample. Moreover, we saw in Lecture 17 that for large enough n, say $n \geq n_0$, $L(n)$ will be very ample, too. Then we have a contradiction because

$$(L^n(-1) \cdot \underline{o}(1)) = -(\underline{o}(1) \cdot \underline{o}(1)) < 0$$

while

$$(L^n(-1) \cdot L(n_0)) = n(L \cdot L) - n_0(\underline{o}(1) \cdot \underline{o}(1)) > 0$$

if n is large enough.

QED

Going back to the examples in Lecture 13, we can check the result. For $\mathbf{P}_1 \times \mathbf{P}_1$, the pairing on the 2-dimensional $\text{Num}(F) \otimes \mathbf{Q}$ is given by the matrix

$$\begin{pmatrix} 0 & 1 \\ 1 & 0 \end{pmatrix}$$

with one positive, one negative eigenvalue. For the second surface, the pairing on the 3-dimensional $\text{Num}(F) \otimes \mathbf{Q}$ is given by the matrix:

$$\begin{pmatrix} -1 & 0 & 0 \\ 0 & -1 & 0 \\ 0 & 0 & 1 \end{pmatrix} .$$

One can picture the situation somewhat like this: take the real vector space $\mathrm{Num}(F) \otimes \mathbf{R}$, and draw in the "light-cone" $(x \cdot x) = 0$. Look at the closure of the set of positive real linear sums of very ample divisor classes:

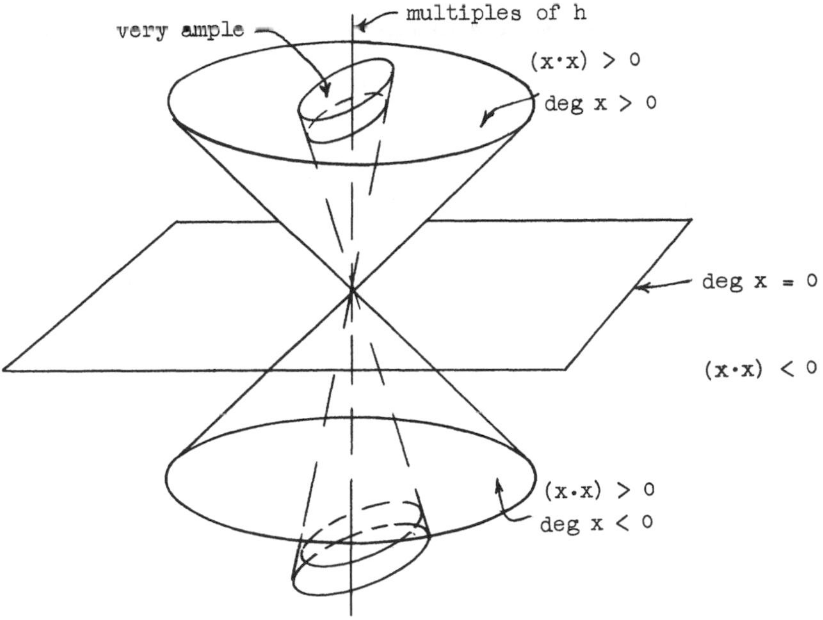

In terms of this diagram, it is useful to look more closely at the numerical criterion for very ampleness in Lecture 17:

$$\deg(L) \geq c_2$$

$$\chi(L) \geq \frac{1-\varepsilon}{2(\underline{o}(1) \cdot \underline{o}(1))} \cdot \deg(L)^2 .$$

Let $\lambda \in \mathrm{Num}(F) \otimes \mathbf{Q}$ be the image of L, and let h be the image of $\underline{o}(1)$. Let Ω be the canonical invertible sheaf on F, and let ω be its image. We use additive notation in $\mathrm{Num}(F)$ for products of invertible sheaves. Then using Proposition 3, Lecture 12, the criterion of Lecture 17 becomes:

a) $\deg(\lambda) = (\lambda \cdot h) \geq c_2$,

b) $(\lambda \cdot \lambda - \omega) + 2\chi(\underline{o}_F) \geq \dfrac{1-\varepsilon}{(h \cdot h)} (\lambda \cdot h)^2$.

In fact, I claim that, with a possible modification of the constants c_2 and ε, b) is implied by the simpler condition:

b') $(\lambda \cdot \lambda) \geq \dfrac{1-\varepsilon}{(h \cdot h)} \cdot (\lambda \cdot h)^2$.

Proof: In fact, let ε' be any positive number smaller than ε, and suppose λ satisfies

$$(*) \quad \begin{cases} \deg(\lambda) \geq c_2 \\ (\lambda \cdot \lambda) \geq \dfrac{1-\varepsilon'}{(h \cdot h)} (\lambda \cdot h)^2 . \end{cases}$$

Then I claim that there is a number A (independent of λ) such that

$$|(\lambda \cdot \omega) - 2\chi(\underline{o}_F)| \leq A \cdot (\lambda \cdot h) \quad .$$

From this it follows immediately that b) holds if $\deg(\lambda)$ is larger than

$$\max \left\{ c_2, \frac{A \cdot (h \cdot h)}{\varepsilon - \varepsilon'} \right\} \quad .$$

To construct A, use the fact that (*) implies $n\lambda$ is represented by a curve for large positive n (the first Proposition of this lecture), and the following easy lemma:

LEMMA: Given any invertible sheaf M on F, there is a constant c_M such that for all curves $D \subset F$,

$$|(\underline{o}_F(D) \cdot M)| \leq c_M \cdot \deg D \quad .$$

Proof: Choose n_0 such that $M(n_0)$ and $M^{-1}(n_0)$ are very ample; then $(\underline{o}_F(D) \cdot M(n_0))$ and $(\underline{o}_F(D) \cdot M^{-1}(n_0))$ are positive, and the lemma follows if $c_M = n_0$.

QED

COROLLARY: There is a positive ε such that if $\lambda \in \mathrm{Num}(F)$ satisfies

a") $\deg(\lambda) > 0$,

b") $(\lambda \cdot \lambda) \geqq \dfrac{1-\varepsilon}{(h \cdot h)} (\lambda \cdot h)^2$,

then all invertible sheaves L representing λ are ample.

Note that these conditions simply define the positive nappe of a cone in $\mathrm{Num}(F) \otimes \mathbb{R}$. On the other hand, conditions a) and b') define the piece of this cone above a certain plane, i.e., a truncated inverted cone. Hence, the set of very ample sheaves includes such a cone.* There is one more result which fits in very nicely with this model. The question arises: what is the exact shape of the real closed cone C_0 spanned by very ample sheaves? It will certainly almost always be bigger than the cone spanned by the points satisfying our numerical criterion. But a theorem of Nakai and Moisezon asserts:

---

* This, at least, makes it quite clear that if L is any invertible sheaf, then $L(n)$ satisfies a) and b) for large enough n.

If L is an invertible sheaf on F, then L is ample if and only if:

a) for all curves $D \subset F$, $(\underline{o}_F(D) \cdot L) > 0$,

b) $(L \cdot L) > 0$,

(cf. Kleiman, Am. J. Math., 1964). In our model, let C be the real closed cone spanned by the invertible sheaves $\underline{o}_F(D)$ for effective D. By the Proposition, this contains the positive numerical cone: $(x, x) \geq 0$, $\deg(x) \geq 0$. Then Nakai's theorem implies that C and C_0 are just dual cones with respect to the intersection pairing!

LECTURE 19

THE PICARD SCHEME : OUTLINE

Our next objective is to prove that the schemes $P(\xi)$ of Lecture 12 exist. Or, equivalently, to prove that there is a universal family of invertible sheaves of numerical type ξ. In this lecture, we shall make some general remarks about the problem, and sketch our method for solving it.

Precisely, the problem is to show that each functor $\underline{\mathrm{Pic}}_F^{\xi}$ is representable. The first thing to notice is that the functors $\underline{\mathrm{Pic}}_F^{\xi}$ are all isomorphic: say ξ_1, ξ_2 are two points in Num(F), and say L_1, L_2 are invertible sheaves on F representing ξ_1 and ξ_2. Define an isomorphism:

$$\underline{\mathrm{Pic}}_F^{\xi_1} \longrightarrow \underline{\mathrm{Pic}}_F^{\xi_2}$$

as follows: given M on $F \times S$ representing an element of $\underline{\mathrm{Pic}}_F^{\xi_1}(S)$, map M to

$$M \otimes p_1^*(L_2 \otimes L_1^{-1}) \quad .$$

This represents an element of $\underline{\mathrm{Pic}}_F^{\xi_2}(S)$ and obviously defines an isomorphism.

The only problem, therefore, is to represent the functor for $\xi = 0$. This functor will be denoted $\underline{\mathrm{Pic}}_F^{\tau}$ (after Grothendieck). This functor is, in a natural way, a group functor: i.e., each of the sets $\underline{\mathrm{Pic}}_F^{\tau}(S)$ is a group and each map between then which is part of the functor, is a homomorphism. Namely, multiply two invertible sheaves on $F \times S$ by tensor product. Therefore, according to the general remarks in Lecture 4, a scheme $P(\tau)$ representing $\underline{\mathrm{Pic}}_F^{\tau}$ is automatically a group scheme. This is essentially Grothendieck's Picard scheme. [Actually, he takes the disjoint union of the schemes representing each $\underline{\mathrm{Pic}}_F^{\xi}$, and calls this the Picard scheme. In the present context, over an algebraically closed field, this is a silly construction: one sees the point only over more complicated base schemes.]

In fact, it will be more convenient to represent $\underline{\mathrm{Pic}}_F^{\xi}$ for one fixed, but very ample ξ. Our method is to choose <u>one</u> ξ which satisfies the numerical criterion of Lecture 17: this guarantees that any L of type

ξ is 0-regular and very ample. Then we shall construct a section s of Φ:

$$\underline{\text{Curves}}_F^{\xi} \underset{s}{\overset{\Phi}{\rightleftarrows}} \underline{\text{Pic}}_F^{\xi} \quad .$$

If Φ <u>admits a section</u> s, then $\underline{\text{Pic}}_F^{\xi}$ <u>is represented by a closed subscheme</u> $P(\xi)$ of $C(\xi)$.

<u>Proof</u>: By hypothesis $\Phi \circ s$ is the identity. On the other hand, $s \circ \Phi$ is a morphism of $\underline{\text{Curves}}_F^{\xi}$ into itself which projects the whole functor onto a subfunctor isomorphic to $\underline{\text{Pic}}_F^{\xi}$. But we know from Lecture 15 that there exists a projective scheme $C(\xi)$ representing $\underline{\text{Curves}}_F^{\xi}$. Therefore $s \circ \Phi$ is indiced by a morphism of schemes:

$$f: \ C(\xi) \to C(\xi) \ .$$

Define $P(\xi)$ as the fibre product in the diagram:

$$\begin{array}{ccc} P(\xi) & \xrightarrow{\ g\ } & C(\xi) \\ \big\downarrow & & \big\downarrow (1,f) \\ C(\xi) & \xrightarrow{\ \Delta\ } & C(\xi) \times C(\xi) \end{array}$$

where Δ is the diagonal morphism.

Then $\mathrm{Hom}(S, P(\xi))$ is isomorphic to the set of pairs $\alpha, \beta \in \mathrm{Hom}(S, C(\xi))$ such that $\Delta(\alpha) = (1, f)(\beta)$, i.e., the points $\alpha \times \alpha$ and $\beta \times f(\beta)$ in $\mathrm{Hom}(S, C(\xi) \times C(\xi))$ are the same. This means that $\mathrm{Hom}(S, P(\xi))$ is isomorphic to the subset of $\mathrm{Hom}(S, C(\xi))$ left fixed by f, i.e., to the subset of $\underline{\text{Curves}}_F^{\xi}(S)$ left fixed by $s \circ \Phi$. Therefore, the functors $h_{P(\xi)}$ and $\underline{\text{Pic}}_F^{\xi}$ are isomorphic.

Finally, since Δ is a closed immersion, the morphism g is a closed immersion so $P(\xi)$ is a closed subscheme of $C(\xi)$.

QED

To construct s, we must do the following: given an invertible sheaf L on $F \times S$, of type ξ along the fibres, construct a relative effective Cartier divisor $D \subset F \times S$ such that

$$\underline{o}_{F\times S}(D) \cong L \otimes p_2^*(M)$$

for some $M \in \mathrm{Pic}(S)$. The construction must have two properties:

(a) if we replace L by $L \otimes p_2^*(M')$ for any $M' \in \mathrm{Pic}(S)$, we should get the <u>same</u> D,

(b) it should commute with base extensions $T \to S$.

The keys to our construction are the following sheaves: given L on $F \times S$, then for any closed point $x \in F$, let i_x: $S \to F \times S$ be the section of p_2

which maps S onto the closed subscheme $\{x\} \times S \subset F \times S$. Then let:

$$M_x = i_x^*(L) \quad .$$

Moreover, let

$$\mathcal{E} = p_{2,*}(L) \ .$$

Then there is a canonical homomorphism

$$h_x\colon \ \mathcal{E} \to M_x$$

for every x; i.e., a section of $\mathcal{E}$ over $U \subset S$ gives a section of L over $F \times U$, hence a section of $i_x^*(L)$ over $U = i_x^{-1}(F \times U)$.

Now recall that ξ was assumed to satisfy the numerical criterion of Lecture 17. Therefore, if an invertible sheaf L' on F is of type ξ, we know that $H^1(F, L') = H^2(F, L') = (0)$, and that L' is very ample. In particular, the restriction of L to any fibre of p_2 is of type ξ. Therefore we know that $\mathcal{E}$ is locally free and that its rank r is determined by ξ alone. Now suppose we choose any $r-1$ closed points $x_1, \ldots, x_{r-1} \in F$. Then we have:

$$\tilde{h} = \sum_{i=1}^{r-1} h_{x_i}\colon \ \mathcal{E} \longrightarrow \bigoplus_{i=1}^{r-1} M_{x_i}$$

hence

$$\wedge\tilde{h}\colon \ \wedge^{r-1} \mathcal{E} \longrightarrow \wedge^{r-1}\left[\bigoplus_{i=1}^{r-1} M_{x_i}\right] \cong \bigotimes_{i=1}^{r-1} M_{x_i} \quad .$$

Dualizing, this gives

$$(\wedge\tilde{h})^*\colon \ \bigotimes_{i=1}^{r-1} M_{x_i}^{-1} \longrightarrow \mathrm{Hom}(\wedge^{r-1} \mathcal{E}, \underline{o}_S) \quad .$$

But

$$\mathrm{Hom}(\wedge^{r-1} \mathcal{E}, \underline{o}_S) \cong \mathcal{E} \otimes (\wedge^r \mathcal{E})^{-1} \quad .$$

[i.e., the canonical pairing of $\wedge^{r-1}(\mathcal{E})$ and $\mathcal{E}$ into the invertible sheaf $\wedge^r \mathcal{E}$ induces a homomorphism from $\mathcal{E}$ to $\underline{\mathrm{Hom}}(\wedge^{r-1} \mathcal{E}, \wedge^r \mathcal{E})$, hence from $\mathcal{E} \otimes \wedge^r \mathcal{E}^{-1}$ to $\underline{\mathrm{Hom}}(\wedge^{r-1} \mathcal{E}, \underline{o}_S)$. It is clear that this is an isomorphism].

Putting all the invertible sheaves together in curly brackets this gives a homomorphism:

$$h'\colon \ \underline{o}_S \to \mathcal{E} \otimes \left\{(\wedge^r \mathcal{E})^{-1} \otimes \left[\bigotimes_{i=1}^{r-1} M_{x_i}\right]\right\}$$

hence a global section:

$$\sigma \in \Gamma\left(F \times S, \ L \otimes p_2^* \left\{(\wedge^r \mathcal{E})^{-1} \otimes \left[\bigotimes_{i=1}^{r-1} M_{x_i}\right]\right\}\right) \quad .$$

Suppose that σ does not vanish identically on any of the fibres of p_2. Then $\sigma = 0$ defines a relative effective Cartier divisor $D \subset F \times S$ such that

$$\underline{o}_{F\times S}(D) \cong L \otimes p_2^*\left\{ (\Lambda^r \mathcal{E})^{-1} \otimes \left[\bigotimes_{i=1}^{r-1} M_{x_i} \right]\right\}$$

which is exactly what we want. Moreover, it is clear that all our steps commute with base extension, and that one winds up with the same D even if you replace L to start with by $L \otimes p_2^*(M)$. Therefore our problem would be solved and s would be constructed, provided only that σ does not vanish identically on any of the fibres of p_2.

What does it mean for σ to vanish identically on the fibre $p_2^{-1}(s)$? Let L_s be the invertible sheaf induced by L on this fibre, and let

$$\varphi_s: F \to \mathbf{P}_{r-1}$$

be the canonical F-valued point of $\mathbf{P}_{r-1}$ defined by L_s [i.e., the one defined by L_s, and $s_1, s_2, \ldots, s_r$, a <u>basis</u> of $H^0(F, L_s)$; cf. Lecture 11].

LEMMA: σ is not identically 0 on $p_2^{-1}(s)$ if and only if $\varphi_s(x_1), \ldots, \varphi_s(x_{r-1})$ span a hyperplane in $\mathbf{P}_{r-1}$.

<u>Proof</u>: Since the construction of σ is functorial, we may as well make the base extension

$$\mathrm{Spec}(k) = \mathrm{Spec}\, \mathcal{K}(s) \to S$$

and replace L by L_s and see whether σ comes out 0 or not. Then $\mathcal{E} = H^0(X, L_s)$ and $M_{x_i} = L_s \otimes \mathcal{K}(x_i)$. Clearly $\sigma \neq 0$ if and only if the $r-1$ linear functionals

$$h_{x_i}: H^0(X, L_s) \to L_s \otimes \mathcal{K}(x_i)$$

are independent, i.e., if the intersection of the kernels of the h_{x_i} is 1-dimensional. But under φ_s^*, $H^0(\mathbf{P}_{r-1}, \underline{o}(1))$ and $H^0(X, L_s)$ are isomorphic. And the linear functional h_{x_i} corresponds to

$$h_{\varphi_x(x_i)}: H^0(\mathbf{P}_{r-1}, \underline{o}(1)) \to \underline{o}(1) \otimes \mathcal{K}(\varphi_s(x_i)) .$$

But an element $h \in H^0(\mathbf{P}_n, \underline{o}(1))$ goes to zero in $\underline{o}(1) \otimes \mathcal{K}(y)$ if and only if the hyperplane defined by h contains y. Therefore the kernel of the h_{x_i}'s is 1-dimensional if and only if there is a unique hyperplane containing the $r-1$ points $\varphi(x_i)$.

QED

Now we know that the image $\varphi_s(F)$ is not contained in any proper linear subspace of $\mathbf{P}_{r-1}$ (cf. Lecture 11). Therefore for almost all $(r-1)$-tuples $x_1, \ldots, x_{r-1}$ of points F, the points $\varphi(x_1), \ldots, \varphi(x_{r-1})$ will be independent and $\sigma \not\equiv 0$ on $p_2^{-1}(s)$. The difficulty, however, is to find <u>one</u> $(r-1)$-tuple which works for every s.

We will not solve this problem: indeed, it may well be that no such $(r-1)$-tuple exists. Instead we shall generalize our method of constructing the section s. We start by choosing a total of $N \cdot r-1$ <u>points on</u> F. We group them into $N-1$ sets of r points, and one set of $r-1$ points:

$$\text{Grouping } (\gamma) \quad \left\{ \begin{array}{l} \{x_{1,1}, x_{1,2}, \ldots, x_{1,r}\} \\ \{x_{2,1}, x_{2,2}, \ldots, x_{2,r}\} \\ \quad \vdots \\ \{x_{N-1,1}, x_{N-1,2}, \ldots, x_{N-1,r}\} \\ \{x_{N,1}, x_{N,2}, \ldots, x_{N,r-1}\} \end{array} \right.$$

For the last $r-1$ points, make the same construction as above, obtaining:

$$\sigma_N \in H^0\Big(F \times S,\ L \otimes p_2^* \Big\{(\wedge^r \mathcal{E})^{-1} \otimes \Big[\bigotimes_{i=1}^{r-1} M_{x_{N,i}}\Big]\Big\}\Big) .$$

For each of the other sets of points, however, we form

$$\widetilde{h} = \sum_{i=1}^{r} h_{x_{k,i}} : \mathcal{E} \to \bigotimes_{i=1}^{r} M_{x_{k,i}} ,$$

hence $\wedge \widetilde{h}$: $\wedge^r \mathcal{E} \to \otimes_{i=1}^{r} M_{x_{k,i}}$. This gives

$$h' : \underline{o}_S \to (\wedge^r \mathcal{E})^{-1} \otimes \Big[\bigotimes_{i=1}^{r} M_{x_{k,i}}\Big]$$

hence a section

$$\sigma_k \in H^0\Big(F \times S,\ p_2^* \Big\{(\wedge^r \mathcal{E})^{-1} \otimes \Big[\bigotimes_{i=1}^{r} M_{x_{k,i}}\Big]\Big\}\Big) .$$

We now put these all together by tensoring to obtain:

$$\sigma = \sigma_1 \otimes \sigma_2 \otimes \ldots \otimes \sigma_N \in H^0\Big(F \times S,\ L \otimes p_2^* \Big\{(\wedge^r \mathcal{E})^{-N} \otimes \Big[\bigotimes_{\text{all } k,\ i} M_{x_{k,i}}\Big]\Big\}\Big)$$

Abbreviate:

$$(\wedge^r \mathcal{E})^{-N} \otimes \Big[\bigotimes_{\text{all } k,i} M_{x_{k,i}}\Big] = K .$$

Now K, up to canonical identifications, is independent of the grouping (γ). Therefore, the result is that for every grouping (γ), we can form a section

$$\sigma_\gamma \in H^0(F \times S,\ L \otimes p_2^*(K)) .$$

Suppose that to each γ we assign a scalar $a_\gamma \in k$. Then we also have the sections $\Sigma_\gamma\, a_\gamma \sigma_\gamma$.

MAIN THEOREM: For a suitable choice of ξ, and of N, and of $N \cdot r - 1$ points on F, and of scalars a_γ, we can reach the result:

$$\left\{ \begin{array}{l} \text{for all invertible sheaves } L \text{ on } F \text{ of type } \xi, \\ \text{the canonical section} \\ \qquad \sum_\gamma a_\gamma \sigma_\gamma \in H^0(F, L) \\ \text{is never zero.} \end{array} \right.$$

When this is proven then, indeed, the sections $\Sigma\, a_\gamma \sigma_\gamma$ always define relative effective Cartier divisors, hence a section s of Φ has been found, and $\mathbf{P}(\xi)$ has been constructed.

LECTURE 20

INDEPENDENT 0-CYCLES ON A SURFACE

In this lecture, we will consider the question of finding finite sets of points on a given surface which are, roughly, "in general position." Fix the surface F, and a very ample invertible sheaf L on F.

1° Definition. A 0-cycle $\mathfrak{A}$ of degree N on F is a formal sum of N (not necessarily distinct) closed points on F:

$$\mathfrak{A} = \sum_{i=1}^{N} P_i \quad .$$

Definition. A 0-cycle $\sum P_i$ is λ-independent if, for all curves $D \subset F$,

$$[\text{Number of } P_i \text{ in } \operatorname{Supp}(D)] \leq \lambda \cdot (\deg D)^2 \quad .$$

First consider the independence of a 0-cycle in the plane: for example, if a 0-cycle is to be 2-independent, then no three points in the cycle should be collinear, no 9 points in the cycle should be on a single conic, etc. This is very weak, of course: there is no reason why even 6 points need be on a single conic. To construct independent 0-cycles by induction on their degree, it is convenient to prove the strongest result:

Proposition 0: For all N, there is a 0-cycle $\mathfrak{A} = \Sigma_{i=1}^{N} P_i$ of degree N on $\mathbf{P}_2$ such that, for all $S \subset \{1, 2, \ldots, N\}$, and for all integers n, if

$$L_{n,S} = \{F \mid F \text{ a homogeneous form in } X_0, X_1, X_2 \text{ of degree } n \text{ such that } F(P_i) = 0, \text{ if } i \in S\} ,$$

then

a) $L_{n,S} = (0)$ if $\operatorname{Card}(S) \geq \dfrac{(n+1)(n+2)}{2}$

b) $\dim L_{n,S} = \dfrac{(n+1)(n+2)}{2} - \operatorname{Card}(S)$ otherwise.

Proof: For $N = 1$, let $\mathfrak{A} = P_1$ be any closed point. Now say $\mathfrak{A} = \Sigma_{i=1}^{N-1} P_i$ is constructed. We must choose P_N so that $\mathfrak{A} + P_N$ meets all requirements: now we need not worry about subsets $S \subset \{1, 2, \ldots, N-1\}$ as they are already taken care of. Say $T \subset \{1, 2, \ldots, N-1\}$ and $S = T \cup \{N\}$. Also, let $L_{n,T}$ and $L_{n,S}$ be the linear spaces defined above. Then the requirements boil down to:

$$L_{n,S} \subsetneq L_{n,T}$$

if $L_{n,T} \neq (0)$, i.e., if $\text{Card}(T) < ((n+1)(n+2))/2$. Namely, by induction, $\dim L_{n,T}$ is given by a) and b); and $L_{n,S}$ has at most codimension 1 in $L_{n,T}$. Therefore, if it is a proper subspace, its dimension is given by a) and b) too.

Let $Z_{n,T}$ be the intersection of the plane curves defined by forms $F \in L_{n,T}$. Then:

$$L_{n,S} \subsetneq L_{n,T} \iff P_N \notin Z_{n,T} \quad .$$

Clearly, if $P_N \in Z_{n,T}$, then the condition $F(P_N) = 0$ is redundant, so $L_{n,S} = L_{n,T}$. But if $P_N \notin Z_{n,T}$, then there is an $F \in L_{n,T}$ such that $F(P_N) \neq 0$; so $F \in L_{n,T} - L_{n,S}$.

Moreover: $Z_{n,T} \supset Z_{n+1,T}$. Namely, let $Q \in Z_{n+1,T}$ and let $F \in L_{n,T}$. Suppose $F(Q) \neq 0$. Let G be a linear form in X_0, X_1, X_2 which is not zero at Q. Then $F \cdot G \in L_{n+1,T}$ and $F \cdot G(Q) \neq 0$ which is a contradiction. Therefore $F(Q) = 0$, and $Z_{n,T} \supset Z_{n+1,T}$. Also, by conditions a) and b) for T, $Z_{n,T} = \mathbf{P}_2$ if and only if $\text{Card}(T) \geq ((n+1)(n+2))/2$.

Putting all this together, the conditions on P_N boil down to:

$$P_N \notin \bigcup_{T \subset \{1,2,\ldots,N-1\}} (Z_{\nu(T),T})$$

where $\nu(T)$ is the least n such that:

$$\frac{(n+1)(n+2)}{2} > \text{Card}(T) \ .$$

Such a P_N obviously exists.

QED

COROLLARY: For all N, there is a 2-independent 0-cycle on $\mathbf{P}_2$ of degree N.

Proof: The $\mathfrak{A}$ just constructed has the property that at most $((n+1)(n+2))/2 - 1$ of its points are on any given curve of degree n. Since

$$\frac{(n+1)(n+2)}{2} - 1 \leq 2 \cdot n^2$$

for all $n \geq 1$, the Corollary follows.

Now consider a general surface F instead of the plane:

Proposition 1: Let F be a non-singular projective surface, and $\underline{o}(1)$ a very ample invertible sheaf. There is a positive λ such that, for all N, there exist λ-independent 0-cycles on F of degree N.

Proof: Let the embedding $F \subset \mathbf{P}_n$ be defined by $\underline{o}(1)$. As in Lecture 16, there is a projection of $\mathbf{P}_n$ onto $\mathbf{P}_2$ which defines a finite, flat morphism

$$\pi : F \to \mathbf{P}_2 .$$

Moreover, we proved in Lecture 16 that if $D \subset F$ is any curve, then

$$\deg D = \deg \pi_*(D) .$$

Let h be the degree of π, i.e. the rank of the locally free sheaf $\pi_*(\underline{o}_F)$. [Actually, $h = (\underline{o}(1)$ [illegible]$)$ but this is irrelevant.] Put $\lambda = 3h$. Given N, $N_0 = [N/h]$, so that $N = h \cdot N_0 + r$, where $0 \leq r \leq h-1$. Choose a 2-independent 0-cycle b on $\mathbf{P}_2$ of degree N_0. Let $\mathfrak{A}' = \pi^*(b)$: how is π^* defined?

Definition: a) $\pi^*(\Sigma_i Q_i) = \Sigma_i \pi^*(Q_i)$,

b) If $Q \in \mathbf{P}_2$ is a closed point, let $\pi^{-1}(Q) = \{P_1, \ldots, P_k\}$, set-theoretically. Then the scheme theoretic fibre is given by:

$$\pi^{-1}(Q) = \mathrm{Spec}\ \{\pi_*(\underline{o}_F)_Q \otimes \mathcal{K}(Q)\} .$$

and $\pi_*(\underline{o}_F)_Q \otimes \mathcal{K}(Q) = \oplus_{i=1}^{k} A_i$, where A_i is an Artin local ring whose Spec is the point P_i.

$$\pi^*(Q) = \sum_{i=1}^{k} \dim_k(A_i) \cdot P_i .$$

Note that the degree of $\pi^*(\mathfrak{A})$ is h times the degree of $\mathfrak{A}$. Finally, let $P_1, \ldots, P_r$ be any r points in F, and let $\mathfrak{A} = \mathfrak{A}' + \Sigma_{i=1}^{r} P_i$. Then I claim that $\mathfrak{A}$ is λ-independent and of degree N. Let $D \subset F$ be any curve:

$$\begin{aligned}
&[\text{Number of points in } \mathfrak{A} \text{ in Supp } D] \\
&\qquad \leq r + [\text{Number of points in } \mathfrak{A}' \text{ in Supp } D] \\
&\qquad \leq h-1 + h \cdot [\text{Number of points in } b \text{ in Supp}(\pi_*(D))] \\
&\qquad \leq h-1 + 2h \cdot \{\deg(\pi_* D)\}^2 \\
&\qquad \leq 3 \cdot h \cdot (\deg D)^2 .
\end{aligned}$$

QED

2° The purpose of this section is to show that λ-independent 0-cycles are good in some other senses too. First introduce a new concept:

Definition: A 0-cycle $\mathfrak{A}$ on $\mathbf{P}_n$ is strongly stable if for all hyperplanes $H \subset \mathbf{P}_n$:

<u>Proposition 2</u>: Let $\mathfrak{A}$ be a strongly stable 0-cycle of degree $k(n+1)$. Then there is a decomposition (γ):

$$\mathfrak{A} = \sum_{i=1}^{k} b_i \quad ,$$

where b_i is 0-cycle of degree $n+1$ consisting of $n+1$ projectively independent points, i.e., points not contained in a hyperplane.

<u>Proof</u>: Look at all decompositions

$$\mathfrak{A} = \sum_{i=1}^{\ell} b_i + \mathfrak{A}'$$

where each of the b_i's consists in $(n+1)$-indepnedent points. Pick one such decomposition such that ℓ is maximal: we want to show that $\ell = k$. Moreover, let L be the linear space spanned by the points in $\mathfrak{A}'$. We shall make a secondary induction on dim L. Clearly, if $L = \mathbf{P}_n$, then one can find $n+1$ independent points in $\mathfrak{A}'$ and form a new cycle $b_{\ell+1}$ out of these, so that ℓ is not maximal. Therefore, $\dim L < n$. Now choose a decomposition such that dim L is maximal among all those with maximal ℓ.

I claim that for some i, $1 \leq i \leq \ell$, the 0-cycle b_i is disjoint from L. If not, then one point of each b_i would be in L. This would give a total of at least $\ell + \deg(\mathfrak{A}')$ points in L. But then

$$\begin{aligned} k &= \frac{\deg \mathfrak{A}}{n+1} \\ &\geq [\text{Number of points in } \mathfrak{A} \text{ in } L] \\ &\geq \ell + \deg(\mathfrak{A}') \\ &= \ell + (k-\ell)(n+1) \\ &> k \, . \end{aligned}$$

This contradiction proves the claim.

Now say b_1 is disjoint from L. Let $b_1 = \Sigma_{i=0}^{n} Q_i$, and let H(i) be the span of all the points $Q_0, \ldots, Q_n$ except Q_i. On the other hand, let $q = \dim L$ and choose $q+1$ points $P_0, P_1, \ldots, P_q$ from $\mathfrak{A}'$ which span L. Let P^* be any point in $\mathfrak{A}'$ other than $P_0, P_1, \ldots,$ or P_q. Since the Q's are independent

$$\bigcap_{i=0}^{n} H(i) = \emptyset \quad .$$

Therefore, there is an i, say i_0, such that $P^* \notin H(i_0)$. Now let

$$b_1^* = \sum_{\substack{i=0 \\ i \neq i_0}}^{n} Q_i + P^*$$

and let $\mathfrak{A}^* = \mathfrak{A}' - P^* + Q_{i_0}$. Since $P^* \notin H(i_0)$, b_1^* still consists of $n+1$

independent points. But now $\mathfrak{A}^*$ contains $P_0, P_1, \ldots, P_q$ and Q_{i_0}. Since b_1 is disjoint from L, $Q_{i_0} \notin L$. Therefore these points span a linear space bigger than L: so $\dim L$ was not maximal.

QED

COROLLARY: Let $\mathfrak{A}$ be a strongly stable 0-cycle of degree $k(n+1)-1$. Then for all closed points $Q \in \mathbf{P}_n$, there is a decomposition (γ):

$$\mathfrak{A} = \sum_{i=1}^{k-1} b_i + b_k^*$$

where $b_1, \ldots, b_{k-1}$ are cycles of $n+1$ independent points, and where b_k^* is a cycle of n independent points spanning a hyperplane H such that $Q \notin H$.

Proof: Apply the Proposition to $\mathfrak{A} + Q$.

The relationship between the two concepts of λ-independence and strong stability is given by:

Proposition 3: Let F be a non-singular projective surface, let $\underline{o}(1)$ be a given very ample sheaf on F, and let $\mathfrak{A}$ be a 0-cycle on F, λ-independent (with respect to $\underline{o}(1)$). Let L be an invertible sheaf on F spanned by its sections and let

$$\varphi: F \to \mathbf{P}_n$$

be the canonical morphism defined by L and its sections. If $\deg(\mathfrak{A}) \geq \lambda(n+1)(\deg L)^2$ then $\varphi_*(\mathfrak{A})$ is a strongly stable 0-cycle on $\mathbf{P}_n$.

Proof: If $H \subset \mathbf{P}_n$ is a hyperplane, then $\varphi^*(H)$ is defined and is a curve in the divisor class of L. Therefore:

$$\begin{aligned}
&[\text{Number of points in } \varphi_*(\mathfrak{A}) \text{ in } H] \\
&\quad \leq [\text{Number of points in } \mathfrak{A} \text{ in } \operatorname{Supp} \varphi^*(H)] \\
&\quad \leq \lambda \cdot \{\deg \varphi^*(H)\}^2 \\
&\quad = \lambda \cdot (\deg L)^2 \\
&\quad \leq \frac{\deg \mathfrak{A}}{n+1} .
\end{aligned}$$

QED

LECTURE 21

THE PICARD SCHEME: CONCLUSION

We can now complete the proof of the existence of the Picard scheme. Recall that we have made a basic choice of a numerical class ξ of invertible sheaves. We shall, at a later point, put more conditions on ξ, but at the moment we know only that the values of $\deg(\xi)$ and $\chi(\xi)$ (defined because of Proposition 3, Lecture 12) satisfy the hypotheses of Theorem 2, Lecture 17. Let λ be an integer such that F admits λ-independent 0-cycles of all degrees. Choose an integer N such that

$$N > \lambda \cdot (\deg \xi)^2 ,$$

and choose a λ-independent 0-cycle $\mathfrak{A}$ on F degree $N \cdot \chi(\xi) - 1$. Now, suppose L is any invertible sheaf of type ξ on F: then

a) $H^1(F, L) = H^2(F, L) = 0$, so that
$\dim H^0(F, L = \chi(L) = \chi(\xi)$,

b) L is very ample.

Let $\varphi\colon F \to P_{r-1}$ be the closed immersion defined by L and its sections, $(r = \chi(\xi))$. Then for all closed points $x \in F$, $\varphi_*(\mathfrak{A} + x)$ is strongly stable by Proposition 3 of the last lecture. And, for all $x \in P_{r-1}$, $\mathfrak{A}$ can be written

$$\mathfrak{A} = \sum_{i=1}^{N-1} b_i + b_N^*$$

such that, for $1 \leq i \leq N-1$, $\varphi_*(b_i)$ consists of r independent points in P_{r-1}, and for $i = N$, $\varphi_*(b_N^*)$ consists of $r-1$ independent points spanning a hyperplane H where $x \notin H$. (Corollary to Proposition 2, Lecture 20.)

Now recall the definitions of Lecture 19: use the $N \cdot r-1$ points of $\mathfrak{A}$, and their grouping (γ) into the b's to define:

$$\sigma_\gamma \in H^0(F, L \otimes_k K)$$

where K is a certain 1-dimensional vector space over k, canonically associated to L and $\mathfrak{A}$ (K is included here not just to be pedantic, but so that the reader does not think anything is being unobtrusively slipped under the table).

Under the above hypotheses, $\sigma_\gamma \neq 0$.

Proof: By definition $\sigma_\gamma = \sigma_1 \otimes \sigma_2 \otimes \ldots \otimes \sigma_N$, where if $1 \leq i \leq N-1$, σ_i is a canonical element of the 1-dimensional vector space

$$K_i = (\Lambda^r H^0(F, L))^{-1} \otimes \left[\bigotimes_{Q \in b_i} M_Q \right]$$

$$M_Q = L \otimes \mathcal{H}(Q) .$$

If $i = N$, then σ_N is a canonical section of $L \otimes_k K_N$, where K_N is the 1-dimensional vector space:

$$K_N = (\Lambda^r H^0(F, L))^{-1} \otimes \left[\bigotimes_{Q \in b_N^*} M_Q \right] .$$

We saw in Lecture 19 that $\sigma_N \neq 0$ if the kernel of all the homomorphisms

$$H^0(F, L) \longrightarrow M_Q$$

for $Q \in b_N^*$, was one-dimensional and if so that σ_N was a non-zero element in the kernel. Moreover, such an element corresponds to $h \in H^0(\mathbf{P}_{r-1}, \underline{o}(1))$ such that $h(\varphi(Q)) = 0$, all $Q \in b_N^*$. But the 0-cycle $\varphi_*(b_N^*)$ spans a hyperplane H not containing x. Therefore, such an h is uniquely determined up to a scalar and $h(x) \neq 0$. Therefore $\sigma_N \neq 0$.

How about the other σ_i's? Going back to the definition, they are not zero if and only if the whole set of r homomorphisms

$$H^0(F, L) \to M_Q$$

for $Q \in b_i$ are independent; for this is equivalent to asking that they induce an isomorphism:

$$\Lambda^r H^0(F, L) \to \bigotimes_{Q \in b_i} M_Q .$$

On the other hand, it is also equivalent to asking that the set of r homomorphisms:

$$H^0(\mathbf{P}_{r-1}, \underline{o}(1)) \to \underline{o}(1) \otimes \mathcal{H}(\varphi(Q))$$

for $Q \in b_i$, are independent. This is true since the 0-cycle $\varphi_*(b_i)$ consists of r independent points.

QED

COROLLARY: For fixed L, but different groupings γ, the elements σ_γ generate $H^0(F, L \otimes_k K)$.

Proof: In the proof just given, the element h can be chosen so that $h(x) \neq 0$ for any $x \in \mathbf{P}_{r-1}$. Therefore, the set of h's which occur span the vector space $H^0(\mathbf{P}_{r-1}, \underline{o}(1))$. Therefore, the set of σ_N's which occur span the vector space $H^0(F, L \otimes_k K_N)$. Therefore the set of σ_γ's which occur span $H^0(F, L \otimes_k K)$.

QED

The obstacle still is that only certain groupings (γ) give rise to non-zero elements in one space $H^0(F, L \otimes_k K)$. Varying L, which (γ) should be chosen? But we have one more gun: we can choose scalars a_γ, one for each grouping γ so that the sum $\Sigma\, a_\gamma \sigma_\gamma$ is not zero for any L. To do this, however, we must look at one very comprehensive family of invertible sheaves of type ξ. One such is gotten as follows: let $D(\xi) \subset F \times C(\xi)$ be the universal family of curves of type ξ. Look at $\mathfrak{L} = \underline{o}_{F\times C(\xi)}(D(\xi))$. This is a family of invertible sheaves over $C(\xi)$ such that every invertible sheaf on F of type ξ appears on one fibre. But the dimension of the base grows with ξ which is awkward. Instead, let ξ_0 be one numerical type satisfying all the same conditions as ξ and let ξ be a much more ample numerical type: in fact, satisfying:

$$(*) \qquad \chi(\xi) > \dim C(\xi_0) \quad .$$

[This can be achieved by first choosing ξ_0, and then letting ξ be $\xi_0 + m \cdot \eta$ for large m, where $\eta \in \mathrm{Num}(F)$ represents $\underline{o}(1)$.] Fix one invertible sheaf M of type $\xi - \xi_0$, and let $D(\xi_0) \subset F \times C(\xi_0)$ be the universal family of curves of type ξ_0. Then

$$\mathfrak{L} = \underline{o}_{F\times C(\xi_0)}\Big(D(\xi_0) \otimes p_1^*(M)\Big)$$

is a family of invertible sheaves of type ξ which also induces every possible sheaf on F of type ξ on some fibre. This is so, because if L is any sheaf of type ξ, then $L \otimes M^{-1}$ is of type ξ_0 so that $H^0(F, L \otimes M^{-1}) \neq (0)$. Therefore $L \otimes M^{-1} \cong \underline{o}_F(D_0)$ and $L \cong \underline{o}_F(D_0) \otimes M$ for some curve D_0. If D_0 defines the closed point $s \in C(\xi_0)$, then L occurs as the sheaf induced by $\mathfrak{L}$ on the fibre over s.

Now, abbreviate $C(\xi_0)$ to S, but let $\mathfrak{L}$ on $F \times S$ still denote the family of sheaves just constructed. Let $\mathcal{E} = p_{2,*}(\mathfrak{L})$. Note that by (*), the rank of $\mathcal{E}$ is bigger than the dimension of S. Let

$$K = (\Lambda^r \mathcal{E})^{-N} \otimes \Big\{ \bigotimes_{Q \in \mathfrak{A}} \mathfrak{M}_Q \Big\}$$

and $\mathfrak{M}_Q = i_Q^*(\mathfrak{L})$. For all groupings (γ) of $\mathfrak{A}$, let

$$\sigma_\gamma \in H^0\Big(F \times S,\ \mathfrak{L} \otimes p_2^*(K)\Big) = H^0(S, \mathcal{E} \otimes K)$$

be the corresponding section.

If the scalars a_γ have the property that for all closed points $s \in S$, the image of the section $\Sigma\, a_\gamma \sigma_\gamma$ in $(\mathcal{E} \otimes K) \otimes \mathfrak{K}(s)$ is not zero, then $\Sigma\, a_\gamma \sigma_\gamma$ meets all the requirements. For the whole construction commutes with base extension, so if L is the sheaf induced by $\mathfrak{L}$ on $p_2^{-1}(s)$, then the image of $\Sigma\, a_\gamma \sigma_\gamma$ is the corresponding $\Sigma\, a_\gamma \sigma_\gamma$ in $H^0(F, L) \otimes K$. And every L occurs over some point s. On the other hand, the sections σ_γ have quite a bit of freedom: for every closed point $s \in S$, the images of the σ_γ generate the vector space $(\mathcal{E} \otimes K) \otimes \mathcal{K}(s)$, (by the Corollary just above). Everything now follows from an easy lemma of Serre:

LEMMA (Serre): Let X be an (algebraic) scheme, and let $\mathcal{E}$ be a locally free sheaf of rank r on X. Let $V \subset H^0(X, \mathcal{E})$ be a finite dimensional vector space and assume:

i) $r > \dim X$,

ii) for all closed points $x \in X$, the map from V to $\mathcal{E} \otimes \mathcal{K}(x)$ is surjective.

Then there is an element $s \in V$ whose image in every space $\mathcal{E} \otimes \mathcal{K}(x)$ is non-zero.

<u>Proof</u>: Let $N = \dim V$ and let $e_1, \ldots, e_N$ be a basis of V. Construct a homomorphism h

$$0 \to \mathcal{N} \xrightarrow{\lambda} \underline{o}_X^N \xrightarrow{h} \mathcal{E} \to 0$$

by $h(a_1, \ldots, a_N) = \sum a_i e_i$. This is surjective by (ii). Let $\mathcal{N}$ be its kernel. Then $\mathcal{N}$ is locally free of rank $N-r$: in fact, tensoring with the residue field $\mathcal{K}(x)$ of any $x \in X$, we obtain:

$$\begin{array}{l} \underline{\mathrm{Tor}}_1^{\underline{o}_X}(\mathcal{E}, \mathcal{K}(x)) \longrightarrow \mathcal{N} \otimes \mathcal{K}(x) \xrightarrow{\lambda_x} \mathcal{K}(x)^N \longrightarrow \mathcal{E} \otimes \mathcal{K}(x) \to 0 \\ \qquad \| \\ \qquad (0) \end{array}$$

and $\underline{\mathrm{Tor}}_1^{\underline{o}_X}(\mathcal{N}, \mathcal{K}(x)) = (0)$.

Pass to the dual exact sequence:

$$0 \to \underline{\mathrm{Hom}}(\mathcal{E}, \underline{o}_X) \longrightarrow \underline{o}_X^N \xrightarrow{\hat{\lambda}} \underline{\mathrm{Hom}}(\mathcal{N}, \underline{o}_X) \to 0 .$$

Then $\hat{\lambda}$ induces (cf. EGA 2, (4.1) and (3.6)) a morphism:

$$\mathbf{P}(\lambda): \quad \mathbf{P}[\underline{\mathrm{Hom}}(\mathcal{N}, \underline{o}_X)] \longrightarrow \mathbf{P}(\underline{o}_X^N) = X \times \mathbf{P}_{N-1} .$$

Now $\mathbf{P}[\underline{\mathrm{Hom}}(\mathcal{N}, \underline{o}_X)]$ is locally a product of X with a projective space of dimension one less than the rank of $\underline{\mathrm{Hom}}(\mathcal{N}, \underline{o}_X)$. Therefore, by hypothesis (i),

$$\dim \mathbf{P}[\underline{\mathrm{Hom}}(\mathcal{N}, \underline{o}_X] = \dim X + N - r - 1 < N - 1 .$$

Look at the composite:

$$p_2 \circ \mathbf{P}(\lambda): \quad \mathbf{P}[\underline{\mathrm{Hom}}(\mathcal{N}, \underline{o}_X)] \to \mathbf{P}_{N-1} .$$

Because the dimension of the domain is less than that of $\mathbf{P}_{N-1}$, it is not surjective. Let $\underline{a} \in \mathbf{P}_{N-1}$ be a closed point outside $\mathrm{Im}(p_2 \circ \mathbf{P}(\lambda))$, and let $a_1, \ldots, a_N$ be homogeneous coordinates of $\underline{a}$. Then I claim that $\sum a_i e_i$ is the sought-for section. Suppose $\sum a_i e_i$ is zero at the closed point

$x \in X$. Then $(a_1,\ldots,a_N)$ is in the sub-vector space $\mathfrak{N} \otimes \mathfrak{K}(x)$ of $\underline{o}_X^N \otimes \mathfrak{K}(x)$, under the inclusion λ_x. Therefore $(a_1,\ldots, a_N)$ defines a linear functional on $\underline{\mathrm{Hom}}(\mathfrak{N}, \underline{o}_X) \otimes \mathfrak{K}(x)$, hence a homomorphism P from the symmetric algebra on $\underline{\mathrm{Hom}}(\mathfrak{N}, \underline{o}_X) \otimes \mathfrak{K}(x)$ to $\mathfrak{K}(x)$. The maximal ideal m_x and the kernel of P define a graded sheaf of ideals in this graded sheaf of algebras: i.e., a point of $\mathbf{P}[\underline{\mathrm{Hom}}(\mathfrak{N}, \underline{o}_X)]$, (cf. Lecture 5, Appendix). It follows immediately that $p_2 \circ \mathbf{P}(\lambda)$ maps this point to $\underline{a}$, which is a contradiction.

QED

LECTURE 22

THE CHARACTERISTIC MAP OF A FAMILY OF CURVES

We are now ready to attack the existence problems A and B raised in Lecture 2. We shall consider first problem B. The first step is to define precisely the "characteristic map" ρ indicated roughly in Lecture 2: this is the fundamental linear estimate for families of curves. First some preliminaries:

(A) We will need the following easy criterion for regularity:

<u>Proposition</u>: Let $\underline{o}$ be a noetherian local ring, and $k \subset \underline{o}$ a subfield isomorphic to the residue field. Then $\underline{o}$ is regular if and only if:

(*) for all finite dimensional local k-algebras A, A_0, and surjective k-homomorphisms $A \to A_0$, the map

$$\mathrm{Hom}_k(\underline{o}, A) \to \mathrm{Hom}_k(\underline{o}, A_0)$$

is surjective.

<u>Proof</u>: The condition that $\underline{o}$ is regular and the condition (*) are both equivalent to the same conditions on the completion $\hat{\underline{o}}$ of $\underline{o}$. Therefore assume $\underline{o}$ is complete, hence by structure theorem on complete local rings, there is a surjective homomorphism

$$k[[X_1, \ldots, X_n]] \xrightarrow{\varphi} \underline{o}$$

Moreover, we can assume that $\varphi X_1, \ldots, \varphi X_n$ induces a <u>basis</u> of m/m^2 ($\underline{m} \subset \underline{o}$). Then if $\underline{o}$ is regular φ is an isomorphism and one easily checks (*) for formal power series rings. Conversely, start with the homomorphism

$$\underline{o} \xrightarrow{\psi_2} \underline{o}/m^2 \xleftarrow{\sim} k[[X_1, \ldots, X_n]]/(X_1, \ldots, X_n)^2 \quad .$$

Lift it via (*) to homomorphisms:

$$\begin{array}{ccc} & \overset{\psi_{m+1}}{\nearrow} & k[[X_1, \ldots, X_n]]/(X_1, \ldots, X_n)^{m+1} \\ & & \downarrow \\ \underline{o} & \xrightarrow[\psi_m]{} & k[[X_1, \ldots, X_n]]/(X_1, \ldots, X_n)^{m} \end{array} \quad .$$

Passing to the limit, one obtains a homomorphism:

$$\underline{o} \xrightarrow{\psi} k[[X_1, \dots, X_n]] \quad .$$

But it is clear that $\psi \circ \varphi$ is an automorphism of $k[[X_1, \dots, X_n]]$, and since φ is surjective, this implies that φ is an isomorphism, i.e., $\underline{o}$ is regular.

QED

(B) Suppose A is a finite dimensional local k-algebra. We will look quite frequently at the schemes $F \times \mathrm{Spec}(A)$, so it seems worthwhile to put together at the outset the basic facts on their structure:

i) As a topological space, $F \times \mathrm{Spec}(A)$ is just F. The only thing changed is the structure sheaf.

ii) $\underline{o}_{F\times \mathrm{Spec}(A)}$ is canonically isomorphic to $\underline{o}_F \otimes_k A$. Namely, notice that the projections $p_1 \colon F \times \mathrm{Spec}(A) \to F$, and $p_2 \colon F \times \mathrm{Spec}(A) \to \mathrm{Spec}(A)$ make $\underline{o}_{F\times \mathrm{Spec}(A)}$ into a sheaf of $\underline{o}_F$-algebras and a sheaf of A-algebras respectively. Therefore, there is a canonical homomorphism:

$$(*) \qquad \underline{o}_F \otimes_k A \to \underline{o}_{F\times \mathrm{Spec}(A)} \quad .$$

But since, for affine open sets $U \subset F$,

$$\Gamma(U, \underline{o}_F \otimes_k A) = \Gamma(U, \underline{o}_F) \otimes_k A$$

and

$$\Gamma(U, \underline{o}_{F\times \mathrm{Spec}(A)}) = \Gamma(U, \underline{o}_F) \otimes_k A \ ,$$

(*) is an isomorphism of sheaves.

iii) Now let $1 = e_1, e_2, \dots, e_n$ be a basis of A over k, where $e_2, \dots, e_n$ span the maximal ideal M. Then

$$\underline{o}_{F\times \mathrm{Spec}(A)} = \underline{o}_F + \sum_{i=2}^{n} e_i \cdot \underline{o}_F$$

and

$$\underline{o}^*_{F\times \mathrm{Spec}(A)} = \underline{o}^*_F + \sum_{i=2}^{n} e_i \cdot \underline{o}_F$$

$$= \underline{o}^*_F \cdot \Big(1 + \sum_{i=2}^{n} e_i \cdot \underline{o}_F\Big) \quad .$$

Moreover, the truncated exponential sequence defines a homomorphism:

$$\Big(\sum_{i=2}^{n} e_i \cdot \underline{o}_F\Big)_+ \longrightarrow \Big(1 + \sum_{i=2}^{n} e_i \cdot \underline{o}_F\Big)_\times \quad .$$

<u>provided</u> $e^p = 0$, <u>all</u> $e \in M$, $p = \mathrm{char}(k)$.

LEMMA: The truncated exponential is always an isomorphism.

Proof: Use the truncated log to get an inverse.

We now come to the main point of this lecture: to investigate the families of curves on F over $\operatorname{Spec} k[\varepsilon]/\varepsilon^2$. We denote $\operatorname{Spec} k[\varepsilon]/\varepsilon^2$ by I. Not only is I a scheme over k, but the augmentation

$$k[\varepsilon]/\varepsilon^2 \to k$$

defines a closed immersion of $\operatorname{Spec}(k)$ into I. In this way, a family of curves over I defines exactly one ordinary curve on F. I itself is like a vector personified: it is a single point with the smallest possible amount of "tangential material" sticking out in one direction. A family of curves over I is basically a curve on F, plus an infinitesimal deformation of this curve.

Fix a curve $D \subset F$.

Definition: $N_D = \underline{o}_D \otimes_{\underline{o}_F} \{\underline{o}_F(D)\}$.

This is an invertible sheaf on D, and if D is non-singular, it can be shown to be the sheaf of germs of sections of the normal bundle. Note the exact sequence:

$$0 \to \underline{o}_F \to \underline{o}_F(D) \to N_D \to 0 \quad .$$

Proposition: There is a natural isomorphism between the set of families of curves $\mathfrak{D} \subset F \times I$, over I, which extend $D \subset F$, and the set of global sections of N_D.

Proof: To define a Cartier divisor $\mathfrak{D} \subset F \times I$ is the same as to give an open covering $\{U_i\}$ of F, and local equations for $\mathfrak{D}$. In view of (B), local equations are of the form:

$$F_i = G_i + \varepsilon \cdot H_i \quad ,$$

where

$$G_i,\ H_i \in \Gamma(U_i, \underline{o}_F) \quad .$$

The induced curve on F itself is defined by the first terms G_i. Assume that this curve is D. Recall that on $U_i \cap U_j$ we must have:

$$F_i = (\text{unit}) \cdot F_j \quad ,$$

or

$$(G_i + \varepsilon H_i) = (a_{ij} + \varepsilon b_{ij}) \cdot (G_j + \varepsilon H_j)$$

where

$$\begin{cases} a_{ij} \in \Gamma(U_i \cap U_j, \underline{o}_F^*) \\ b_{ij} \in \Gamma(U_i \cap U_j, \underline{o}_F) \quad . \end{cases}$$

This gives the equations:

$$G_i = a_{ij} \cdot G_j$$
$$H_i = a_{ij}H_j + b_{ij}G_j$$

hence

$$\frac{H_i}{G_i} - \frac{H_j}{G_j} = b_{ij} \cdot a_{ji} \quad .$$

But since G_i is a local equation for D, H_i/G_i is a section of $\underline{o}_F(D)$, and these equations say that $\{H_i/G_i\}$ patch together as sections of N_D. This is the section corresponding to $\mathfrak{D}$.

Now suppose that with respect to some open covering $\{U_i\}$, two sets of local equations F_i, F_i' gave the same sections of N_D. Then

$$\frac{H_i}{G_i} - \frac{H_i'}{G_i'} = c_i \in \Gamma(U_i, \underline{o}_F) \quad .$$

Also, since G_i and G_i' are both local equations for D, G_i/G_i' is a unit d_i in U_i. Then it follows that

$$(G_i + \varepsilon H_i) = (d_i + \varepsilon c_i \cdot d_i) \cdot (G_i' + \varepsilon H_i')$$

hence the two divisors $\mathfrak{D}$ and $\mathfrak{D}'$ are equal. Finally, it is easy to check that every section of N_D defines a divisor $\mathfrak{D}$ extending D in this way.

QED

COROLLARY 1: Given a family of curves $\mathfrak{D} \subset F \times S$, and a closed point $s \in S$, there is a canonical linear homomorphism

$$\rho: \quad \left\{ \begin{array}{l} \text{the Zariski tangent} \\ \text{space } T_s \text{ to } S \text{ at } s \end{array} \right\} \to H^0(F, N_{D_s})$$

(where $D_s \subset F$ is the curve induced by $\mathfrak{D}$). This is the <u>characteristic map</u> of the family.

<u>Proof</u>: given $t \in T_s$, we have a canonical

$$f: \ I \to S$$

with image s (cf. Lecture 4, Appendix). Then, by base extension f, we obtain a family of curves $\mathfrak{D}_f \subset F \times I$ which extends D_s. By the proposition, $\mathfrak{D}_f$ corresponds to an element $\rho(t) \in H^0(F, N_{D_s})$. To show that ρ is linear, use the functorial characterization of the vector space structure on T_s (Appendix, Lecture 4), and check that this agrees with structure we have introduced directly.

COROLLARY 2: For the universal family of curves $\mathfrak{D} \subset F \times C(\xi)$, ρ is an isomorphism at all closed points $s \in C(\xi)$.

Proof: Following the proof of the previous corollary, the set of t is always isomorphic to the set of f; and the set of $\alpha \in H^0(F, N_{D_s})$ is isomorphic by the proposition to the set of families $\mathfrak{D}' \subset F \times I$ extending D_s. But by definition of a universal family, every $\mathfrak{D}'$ equals a $\mathfrak{D}_f$ for a unique f, so the set of $\mathfrak{D}'$ and the set of f are isomorphic too.

QED

This would appear to answer the fundamental Problem B of Lecture 2. But in fact, it does not. We have only generalized the concept of a family of curves from the intuitive one where the base is a non-singular variety, to a "phony" one where the Zariski tangent space to the base can be huge, but the base can still be only one point! The burden of the problem of really constructing families of curves is shifted to the question of ascertaining whether the universal base is reduced, or (better) non-singular.

Example: The following is due to Severi and Zappa: let C be an elliptic curve over k, and consider vector bundles $\mathcal{E}$ of rank 2 over C which fit into exact sequences:

$$0 \to \underline{o}_C \to \mathcal{E} \to \underline{o}_C \to 0 .$$

By the general theory of sheaves, such extensions are classified by elements of:

$$\mathrm{Ext}^1_{\underline{o}_C}(\underline{o}_C, \underline{o}_C) \cong H^1(C, \underline{o}_C) .$$

But $H^1(C, \underline{o}_C)$ is a 1-dimensional vector space; let $\mathcal{E}$ correspond to a non-zero element. We take $F = \mathbf{P}(\mathcal{E})$, (cf. Lecture 5). This is a ruled surface, i.e., there is a canonical projection

$$\pi: F \to C$$

making F into a bundle over C with fibre $\mathbf{P}_1$. We can be very explicit: let P, Q be two distinct points on C. Up to adding a constant and multiplying by a scalar, there is a unique function f on C with simple poles at P and Q, and no other poles. The covering

$$\begin{aligned} C &= (C - P) \cup (C - Q) \\ &= U_P \cup U_Q \end{aligned}$$

and

$$f \in \Gamma(U_P \cap U_Q, \underline{o}_C)$$

give a 1-Czech co-cycle on C which represents the generator of $H^1(C, \underline{o}_C)$ (up to a scalar). Then one can check that

$$F = [\mathbf{P}_1 \times U_P] \cup [\mathbf{P}_1 \times U_Q]$$

and that if t_P is a coordinate on $\mathbf{P}_1$ in the first patch, t_Q one in the second, then the patching identifies the closed points

$$(t_P, x) \in \mathbf{P}_1 \times U_P$$
$$(t_Q, x) \in \mathbf{P}_1 \times U_Q$$

when $x \in U_P \cap U_Q$, $t_P - t_Q = f$.

Now the curves given by $(\infty) \times U_P$ and $(\infty) \times U_Q$ coincide over $U_P \cap U_Q$: i.e., the first has local equation t_P^{-1}, the second has local equation t_Q^{-1}, and

(#) $$t_P^{-1}/t_Q^{-1} = 1 - f \cdot t_P^{-1}, \text{ a unit in a neighborhood of } (\infty) \times (U_P \cap U_Q) .$$

Call this curve E. E is a section of the morphism π, and therefore is an irreducible non-singular curve on F, isomorphic to C. Moreover,

$$\underline{o}_F(E) \cong t_P \cdot \underline{o}_F \quad \text{in } \mathbf{P}_1 \times U_P$$
$$\cong t_Q \cdot \underline{o}_F \quad \text{in } \mathbf{P}_1 \times U_Q .$$

Therefore, $N_E \cong \underline{o}_E$ in $E \cap (\mathbf{P}_1 \times U_P)$
$\cong \underline{o}_E$ in $E \cap (\mathbf{P}_1 \times U_Q)$

and the patching on the intersection is defined by the restriction to E of t_P^{-1}/t_Q^{-1}. By (#), this is 1, hence $N_E \cong \underline{o}_E$ globally on E. Therefore:

$$H^0(F, N_E) \cong H^0(E, \underline{o}_E) \cong k .$$

This means that the universal family $C_F^{d,1}$ of curves of degree d, genus 1 containing E has a non-trivial Zariski-tangent space at the point e corresponding to E.

On the other hand, it is easy to check that e alone is a <u>component</u> of $C_F^{d,1}$. For one can show that if a second curve $E' \subset F$ corresponded to a point e' in the <u>same</u> component of $C_F^{d,1}$ as e, then $E \cap E' = \emptyset$. [It would follow that the <u>sheaf</u> $\underline{o}_F(E')$ was a deformation of the sheaf $\underline{o}_F(E)$, hence $\underline{o}_E \otimes \underline{o}_F(E')$ would be a deformation, on E, of N_E; but the former has a section which vanishes at $E \cap E'$, and N_E has a section which is nowhere zero; since their Euler characteristics are the same, this means that $E \cap E' = \emptyset$.] But also the degree of E' over C must be 1 like that of E over C: therefore E' would also be a section of π and would have local equations:

$$t_P = g_P(x) \quad \text{in } \pi^{-1}(U_P), \; g_P \in \Gamma(U_P, \underline{o}_C)$$
$$t_Q = g_Q(x) \quad \text{in } \pi^{-1}(U_Q), \; g_Q \in \Gamma(U_Q, \underline{o}_C) .$$

Then $g_P - g_Q = f$, and f is a Czech co-boundary which is a contradiction.

LECTURE 23

THE FUNDAMENTAL THEOREM VIA KODAIRA-SPENCER

We are now ready to prove the theorem announced in Lecture 2, for which two analytic proofs were sketched. We will prove the strongest known form of this result in the form B given at that time.

<u>Definition</u>: A curve $D \subset F$ is <u>semi-regular</u> if

$$H^1(\underline{o}_F(D)) \to H^1(N_D) \text{ is the zero-map .}$$

THEOREM: (Severi-Kodaira-Spencer). Let $D_0 \subset F$ be a curve of type ξ. Let D_0 correspond to the closed point $\delta \in C(\xi)$. If

a) $\text{char}(k) = 0$,

b) D_0 is semi-regular,

then $C(\xi)$ is non-singular at δ.

<u>Proof</u>: We shall use the criterion of section (A), Lecture 22. Let A be a finite dimensional local k-algebra, $I \subset A$ an ideal and $\overline{A} = A/I$. We must show that every curve $\overline{D} \subset F \times \text{Spec}(\overline{A})$ which extends D_0 also extends to a curve $D \subset F \times \text{Spec}(A)$. Clearly we can also assume that $\dim I = 1$, and let $I = \eta \cdot A$. Fix local equations $\overline{F}_i$ of $\overline{D}$ in some affine open covering $\{U_i\}$ of F. To start with, lift $\overline{F}_i$ arbitrarily to elements

$$F_i \in \Gamma(U_i, \underline{o}_F \otimes_k A) \quad .$$

The trouble is that these do not define a curve D unless F_i <u>and</u> F_j differ by a <u>unit</u> <u>in</u> $U_i \cap U_j$. But, in any case, there are units $\overline{G}_{ij}$ on $U_i \cap U_j$ in $(\underline{o}_F \otimes \overline{A})^*$ such that:

$$\overline{F}_i = \overline{G}_{ij} \cdot \overline{F}_j \quad .$$

Lift $\overline{G}_{ij}$ arbitrarily to $G_{ij} \in \Gamma(U_i \cap U_j, (\underline{o}_F \otimes A)^*)$. Then

$$F_i - G_{ij} \cdot F_j = \eta \cdot h_{ij}, \; h_{ij} \in \Gamma(U_i \cap U_j, \underline{o}_F)$$

and we must show that for a suitable choice of F_i and G_{ij} we can make all the h_{ij} equal to 0. First note the identity:

$$\begin{aligned}\eta(h_{ij} + G_{ij} \cdot h_{jk}) &= F_i - G_{ij}F_j + G_{ij}(F_j - G_{jk}F_k) \\ &= F_i - G_{ij}G_{jk}F_k \\ &= \eta \cdot h_{ik} + (G_{ik} - G_{ij}G_{jk})F_k \ .\end{aligned}$$

Let $G_{ij}^{(0)}$ and $F_k^{(0)}$ denote the images of G_{ij} and F_k in $\underline{o}_F$. Then we get:

$$h_{ij} + G_{ij}^{(0)} \cdot h_{jk} = h_{ik} + \left(\frac{G_{ik} - G_{ij}G_{jk}}{\eta} \right) \cdot F_k^{(0)} \ .$$

Since $F_k^{(0)}$ is a local equation for D_0, and $F_i^{(0)} = G_{ij}^{(0)} \cdot F_j^{(0)}$, this gives:

$$(\dagger) \qquad \frac{h_{ij}}{F_i^{(0)}} + \frac{h_{jk}}{F_j^{(0)}} = \frac{h_{ik}}{F_i^{(0)}} + \left[\frac{1 - G_{ij}G_{jk}G_{ik}^{-1}}{\eta} \right]$$

hence

$$\left\{ \frac{h_{ij}}{F_i^{(0)}} \right\}_{\text{all } i,j}$$

is a 1-Czech co-cycle for the sheaf N_D. Let this correspond to $\overline{\mathfrak{H}} \in H^1(N_D)$.

$\mathfrak{H}$ is the obstruction to finding D! Let's check that if $\overline{\mathfrak{H}} = 0$, then D exists. In fact, suppose we make the changes:

$$\begin{aligned} F_i' &= F_i + \eta f_i \ , \\ G_{ij}' &= G_{ij} + \eta \cdot g_{ij} \ . \end{aligned}$$

Then one computes:

$$\begin{aligned}\eta \cdot h_{ij}' &= F_i' - G_{ij}' \cdot F_j' \\ &= F_i - G_{ij}F_j + \eta \cdot f_i - \eta f_j \cdot G_{ij} - \eta F_j g_{ij} \\ &= \eta(h_{ij} + f_i - f_j G_{ij} - F_j g_{ij}) \ .\end{aligned}$$

Since g_{ij} is an arbitrary element of $\Gamma(U_i \cap U_j, \underline{o}_F)$, we can make h_{ij}' equal to 0 for all i, j if we can make

$$h_{ij} + f_i - f_j G_{ij}^{(0)} \in (F_j^{(0)}) \ ,$$

by a suitable choice of $\{f_i\}$. But this means that

$$-\frac{h_{ij}}{F_i^{(0)}} \equiv \frac{f_i}{F_i^{(0)}} - \frac{f_j}{F_j^{(0)}} \pmod{\underline{o}_F} \ ,$$

or that $-\mathfrak{H}$ is a Czech co-boundary in the sheaf N_D. This proves that D exists if $\mathfrak{H} = 0$.

Now by hypothesis b), the homomorphism

$$H^1(N_D) \xrightarrow{\partial} H^2(\underline{o}_F)$$

coming from the exact sequence

$$0 \to \underline{o}_F \to \underline{o}_F(D) \to N_D \to 0$$

is injective. Therefore it suffices to prove that $\partial(\mathfrak{H}) = 0$. But since the sections $h_{ij}/F_i^{(0)}$ of $\underline{o}_F(D)$ lift the co-chain representing $\mathfrak{H}$ into $\underline{o}_F(D)$, it follows from formula (†) that $\partial(\mathfrak{H})$ is represented by the Czech 2 co-cycle:

$$\sigma_{ijk} = \left[\frac{1 - G_{ij} \cdot G_{jk} \cdot G_{ik}^{-1}}{\eta} \right] .$$

But $\{\sigma_{ijk}\}$ is an obstruction to lifting the 1-co-cycle in $\{\overline{G_{ij}}\}$ in $(\underline{o}_F \otimes \overline{A})^*$ to a co-cycle in $(\underline{o}_F \otimes A)^*$: for if it can be lifted, then we may choose $\{G_{ij}\}$ such that $G_{ij} \cdot G_{jk} = G_{ik}$, i.e., $\sigma_{ijk} = 0$. Everything follows now from:

LEMMA: $(\underline{o}_F \otimes A)^* \to (\underline{o}_F \otimes \overline{A})^* \to 1$ splits.

Proof: One merely uses the exponential, as the characteristic is 0:

$$\begin{array}{ccc}
(\underline{o}_F \otimes A)^* & \longrightarrow & (\underline{o}_F \otimes \overline{A})^* \\
\wr\Vert & & \wr\Vert \\
\underline{o}_F^* \cdot (1 + \underline{o}_F \otimes M) & \longrightarrow & \underline{o}_F^* \cdot (1 + \underline{o}_F \otimes \overline{M}) \\
\wr\Vert \ \exp & & \wr\Vert \ \exp \\
\underline{o}_F^* \cdot (\underline{o}_F \otimes M)_+ & \longrightarrow & \underline{o}_F^* \cdot (\underline{o}_F \otimes \overline{M})_+
\end{array}$$

Now since $M \to \overline{M}$ splits as a surjection of vector spaces, $\underline{o}_F \otimes M \to \underline{o}_F \otimes \overline{M}$ splits as a surjection of sheaves of abelian groups. This proves the lemma.

QED

COROLLARY: Let $D \subset F$ satisfy the hypotheses of the theorem. Then δ is contained in only one component Z of $C(\xi)$ and

$$\dim Z = \dim H^0(F, N_D).$$

Proof: Since the local ring $\underline{o}_\delta$ of $C(\xi)$ at δ is regular

$$\dim Z = \dim \underline{o}_\delta = \dim T_\delta = \dim H^0(F, N_D)$$

by Corollary 2 of Lecture 22.

To properly understand this theorem, it should be added that the requirement of semi-regularity is very weak. Of course, it must be violated by the quite pathological curve E in the example of Lecture 22; but a 1-regular curve is a semi-regular, and we know that for every invertible sheaf L on F, there is an m_0 such that all curves with global equations in $H^0(L(m))$ are 1-regular if $m \geq m_0$. Looking back at the examples of Lecture 1, it will be seen that all the curves not described as superabundant are 1-regular, hence semi-regular. Moreover, look at the analogous case where F is replaced by a non-singular curve γ and $C(\xi)$ is replaced by

C(d)—the universal family of 0-cycles on γ of degree d. Then every 0-cycle $D \subset \gamma$ is semi-regular since N_D has 0-dimensional support, hence

$$H^1(N_D) = (0) .$$

In fact, as is well-known, C(d) is just the d-th symmetric power of γ which is non-singular.

On the other hand, the requirement of characteristic 0 is quite central. For the last four lectures we shall try to get closer to the heart of the theorem so as to bring out several ways in which the characteristic restriction can be "explained." To see what should come next, note that what the proof really does is to reduce the lifting problem for $\overline{D}$ to the problem for its associated invertible sheaf $\underline{o}_{F\times \mathrm{Spec}(\overline{A})}(\overline{D})$ defined by the co-cycle $\overline{G}_{ij}$. Then why not eliminate $\overline{D}$ entirely from the problem, and prove the theorem in form A of Lecture 2-entirely in terms of invertible sheaves?

LECTURE 24

THE STRUCTURE OF Φ

1° In this lecture we want to put together our whole set-up: in Lecture 15, we constructed the schemes $C(\xi)$ parametrizing curves; in Lecture 21, we constructed the schemes $P(\xi)$ parametrizing invertible sheaves. The morphism of functors

$$D \mapsto \underline{o}(D)$$

induces a fundamental morphism of schemes

$$\Phi: \quad C(\xi) \rightarrow P(\xi) \quad .$$

In Lecture 13 we described the fibre functors $\underline{\text{Lin}}\ \underline{\text{Sys}}_L$: now that we have represented $\underline{\text{Curves}}_F$ and $\underline{\text{Pic}}_F$ we get the Corollary:

COROLLARY: <u>The fibres of</u> Φ <u>are projective spaces</u>. <u>In fact</u>, <u>if the sheaf</u> L <u>on</u> F <u>corresponds to</u> $\lambda \in P(\xi)$, <u>then canonically</u>:

$$\Phi^{-1}(\lambda) \cong \mathbf{P}[\widehat{H^0(L)}] \quad .$$

The global structure of Φ can be described somewhat similarly (cf. Grothendieck's Bourbaki talk, exposé 232, p. 11). The interesting thing is that, for different ξ's, the schemes $P(\xi)$ are all isomorphic, whereas the schemes $C(\xi)$ over them are very different—for $\deg(\xi) < 0$, they are empty; for $\deg(\xi) \rightarrow +\infty$, they increase indefinitely in dimension. For some ξ, Φ is a fairly complicated fibering, and its explicit description requires some technical concepts coming out in the further development of the theory of 3°, Lecture 7. Therefore, we only give the result in a special case.

Let $U \subset P(\xi)$ be given such that:

(*) for all closed points $x \in U$, if L_x is the invertible sheaf on F corresponding to x, then $H^1(F, L_x) = (0)$.

For example, if $D \subset F$ is a curve for which $H^1(F, \underline{o}_F(D)) = (0)$, and if D corresponds to the point $\delta \in C(\xi)$, then some neighborhood U of $\Phi(\delta) \in P(\xi)$ satisfies (*) (in virtue of the results of 3°, Lecture 7).

<u>Proposition</u>: There is a locally free sheaf $\mathcal{E}$ on U, and a commutative diagram:

$$\begin{array}{ccccc} C(\xi) & \supset & \Phi^{-1}(U) & \cong & \mathbf{P}(\mathcal{E}) \\ {\scriptstyle \Phi}\downarrow & & \downarrow & \swarrow {\scriptstyle \pi} & \\ P(\xi) & \supset & U & & \end{array}$$

<u>Proof</u>: Let L be the universal family of invertible sheaves on $F \times U$. Abbreviate $p_2\colon F \times U \to U$ to p. According to Lecture 7, 3°, $p_*(L)$ is a locally free sheaf on U, and if $g\colon T \to U$ is any morphism and if one makes the base extension:

$$\begin{array}{ccc} F \times T & \xrightarrow{\ h\ } & F \times U \\ {\scriptstyle q}\downarrow & & \downarrow{\scriptstyle p} \\ T & \xrightarrow{\ g\ } & U \end{array}$$

then $q_*(h^*(L)) \cong g^*(p_*(L))$. Now let $\mathcal{E}$ be the dual locally free sheaf to $p_*(L)$, i.e.,

$$\mathcal{E} = \underline{\mathrm{Hom}}_{\underline{o}_U}(p_*(L), \underline{o}_U) \quad .$$

We shall now prove that $\Phi^{-1}(U)$ and $\mathbf{P}(\mathcal{E})$ are isomorphic over U by the same method used in Lecture 13 to show that $\underline{\mathrm{Lin}}\ \underline{\mathrm{Sys}}_L$ is represented by projective space: we shall give an isomorphism between their functors of points. More precisely, given a T-valued point $g\colon T \to U$ of U, we shall give a natural isomorphism between the set of T-valued points of $\Phi^{-1}(U)$ over g and the set of T-valued points of $\mathbf{P}(\mathcal{E})$ over g. Since this isomorphism will be functorial in g, the theorem will be proven. But proceed in several stages:

(i) $\left\{\begin{array}{l}\text{set of } T\text{-valued points} \\ \text{of } \Phi^{-1}(U) \text{ over } g\end{array}\right\} \cong \left\{\begin{array}{l}\text{set of families of curves } D \subset F \times T \\ \text{such that, for some invertible sheaf} \\ M \text{ on } T,\ \underline{o}_{F\times T}(D) \cong h^*(L) \otimes q^*(M)\end{array}\right\}$.

This follows by the functorial definition of $C(\xi)$ and of Φ.

(ii) $\left\{\begin{array}{l}\text{set of families of curves} \\ D \subset F \times T \text{ such that, for} \\ \text{some invertible sheaf } M \\ \text{on } T, \\ \underline{o}_{F\times T}(D) \cong h^*(L) \otimes q^*(M)\end{array}\right\} \cong \left\{\begin{array}{l}\text{set of invertible sheaves } M \text{ on} \\ T, \text{ and sections of } h^*(L) \otimes q^*(M) \\ \text{inducing non-zero sections in each} \\ \text{fibre } F \times \{t\} \text{ over } T.\end{array}\right\}$

This follows because D is just a relative Cartier divisor over T whose global equation is a section of a sheaf of the form $h^*(L) \otimes q^*(M)$; and an arbitrary Cartier divisor on $F \times T$ is a <u>relative</u> Cartier divisor if its global equation is a non-zero divisor in each fibre over T, i.e., if it is non-zero there.

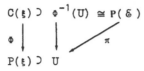

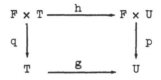

But a section σ of $h^*(L) \otimes q^*(M)$ over $F \times T$ is the same thing as a section τ over T of

$$q_*(h^*(L) \otimes q^*(M)) \cong q_*h^*(L) \otimes M$$
$$\cong g^*(p_*(L)) \otimes M .$$

Moreover, the condition that σ should induce non-zero sections on each fibre over T is the same as the condition that τ should have a non-zero image in

$$\{g^*[p_*(L)] \otimes M\} \otimes \mathcal{K}(t)$$

for all closed points $t \in T$. But a section τ of $g^*[p_*L] \otimes M$ is the same thing as a homomorphism h:

$$\underline{\mathrm{Hom}}_{\underline{o}_T} \{g^*(p_*L), \underline{o}_T\} \xrightarrow{h} M$$

i.e., given a homomorphism from $g^*(p_*L)$ to $\underline{o}_T$ and a section of $g^*[p_*L] \otimes M$, one gets a section of M: this is an h. Moreover, the condition on τ is equivalent to the condition that h be surjective. Finally, since

$$\underline{\mathrm{Hom}}_{\underline{o}_T} (g^*(p_*L), \underline{o}_T) \cong g^*[\underline{\mathrm{Hom}}_{\underline{o}_U} (p_*L, \underline{o}_U)]$$
$$\cong g^*\mathcal{E}$$

we get:

$$\text{(iii)} \quad \left\{ \begin{array}{l} \text{set of invertible sheaves } M \text{ on } T, \\ \text{and sections of } h^*(L) \otimes q^*(M) \text{ in-} \\ \text{ducing non-zero sections in each} \\ \text{fibre } F \times \{t\} \text{ over } T. \end{array} \right\} \cong \left\{ \begin{array}{l} \text{set of invertible sheaves} \\ M \text{ on } T, \text{ and surjections} \\ \quad \Phi\colon\ g^*(\mathcal{E}) \to M . \end{array} \right\}$$

But by the Appendix to Lecture 5, this latter set is isomorphic to the set of T-valued points of $P(\mathcal{E})$ lifting the given T-valued point g of U. This gives the sought-for isomorphism.

QED

2° Next we want to describe the infinitesimal structure of $P(\xi)$, i.e., its I-valued points, just as we have described those of $C(\xi)$ —intrinsically on F. We may as well look at the case $\xi = 0$: this is the scheme we called $P(\tau)$ before. $P(\tau)$ is a group scheme, and consequently homogeneous in the following sense: if x, y are two closed points of $P(\tau)$, there is an automorphism T of $P(\tau)$ such that $T(x) = y$. This, in itself, immetidately implies that all topological components of $P(\tau)$ are irreducible; that they are all isomorphic to each other; that they have no embedded components; and that $P(\tau)_{red}$ is non-singular. [The last by homogeneity and the fact that there is an open dense subset $U \subset P(\tau)_{red}$ which is non-singular—cf. Lecture 11, (V).] In fact, $P(\tau)_{red}$ is easily checked to be a group scheme itself, using Remark (V) of Lecture 11. Also the component of $P(\tau)_{red}$

containing the identity e is a group scheme: this is the classical <u>Picard variety of</u> F.

Because $P(\tau)$ is a commutative group scheme, for any x, y there is even a canonical automorphism T such that $T(x) = y$. In particular, these automorphisms give canonical isomorphisms of the Zariski tangent spaces at all the closed points of $P(\tau)$ with each other. Therefore, we may limit ourselves to considering the I-valued points of $P(\tau)$ whose underlying k-valued point is the identity 0. Use the truncated exponential sequence:

$$0 \to \underline{o}_F \xrightarrow{\alpha} \underline{o}^*_{F\times I} \underset{\leftarrow - - -}{\longrightarrow} \underline{o}^*_F \longrightarrow 0$$

where $\alpha(f) = 1 + \varepsilon \cdot f$ (cf. Lecture 22, (B)). This splits since $\underline{o}_{F\times I}$ is also a sheaf of $\underline{o}_F$-algebras via the projection $p_1\colon F \times I \to F$. This gives the diagram of groups:

$$\begin{array}{ccccccc}
0 \to H^1(\underline{o}_F) & \longrightarrow & H^1(\underline{o}^*_{F\times I}) & \underset{\leftarrow - - - - - -}{\longrightarrow} & H^1(\underline{o}^*_F) & \longrightarrow & 0 \\
 & & & & \parallel & & \\
\wr\Vert & & \wr\Vert & & \underline{\mathrm{Pic}}(F) & & \\
 & & & & \wr\Vert & & \\
0 \to \begin{bmatrix}\text{Group of I-valued}\\ \text{pts. at } 0 \in P(\tau)\end{bmatrix} & \longrightarrow & \begin{bmatrix}\text{Group of I-valued}\\ \text{pts. of } \coprod P(\xi)\end{bmatrix} & \longrightarrow & \begin{bmatrix}\text{Group of k-valued}\\ \text{pts. of } \coprod P(\xi)\end{bmatrix} & \to & 0
\end{array}$$

In other words, the Zariski tangent space T_0 at the identity is canonically isomorphic to $H^1(F, \underline{o}_F)$. One must check that this is actually an isomorphism of vector spaces. This is left to the reader: it can be done via the methods of the Appendix to Lecture 4.

3° Now suppose that δ is a closed point of $C(\xi)$. Let $\lambda = \Phi(\delta)$. The morphism Φ induces an exact sequence of vector spaces:

$$(\#) \qquad 0 \to \begin{bmatrix}\text{Zariski tangent}\\ \text{space to fibre}\\ \Phi^{-1}(\lambda) \text{ at } \delta\end{bmatrix} \to \begin{bmatrix}\text{Zariski tangent}\\ \text{space to } C(\xi)\\ \text{at } \delta\end{bmatrix} \xrightarrow{\Phi_*} \begin{bmatrix}\text{Zariski tangent}\\ \text{space to } P(\xi)\\ \text{at } \lambda\ .\end{bmatrix}$$

We want to interpret this whole sequence intrinsically on F. But look at the exact sequence of sheaves:

$$0 \to \underline{o}_F \to \underline{o}_F(D) \to N_D \to 0$$

where $D \subset F$ is the curve corresponding to δ. This defines the exact sequence of vector spaces:

$$(\#)' \qquad 0 \to \frac{H^0(\underline{o}_F(D))}{H^0(\underline{o}_F)} \to H^0(N_D) \xrightarrow{\partial} H^1(\underline{o}_F)$$

a) $H^0(N_D) \cong \begin{bmatrix}\text{Zariski tangent} \\ \text{space to } C(\xi) \text{ at } \delta\end{bmatrix}$ by Lecture 22

b) $H^1(\underline{o}_F) \cong \begin{bmatrix}\text{Zariski tangent} \\ \text{space to } P(\xi) \text{ at } \lambda\end{bmatrix}$ by 2°,

and by the automorphism T of $\coprod_\xi P(\xi)$ taking 0 to λ

i.e., T is translation by λ.

Proposition: The homomorphisms Φ_* and ∂ in (#) and (#)' are the same under these identifications of the vector spaces.

Check of compatibility: Let D be defined by local equations G_i in affine open sets $\{U_i\}$. Any section of N_D is defined by data:

$$H_i/G_i, \quad G_i, H_i \in \Gamma(U_i, \underline{o}_F)$$

where

$$H_i/G_i - H_j/G_j \in \Gamma(U_i \cap U_j, \underline{o}_F) \quad ,$$

and the corresponding curve $\overline{\mathfrak{D}}$ in $F \times I$ is given by local equations:

$$F_i = G_i + \varepsilon H_i \quad .$$

Then the invertible sheaf $\Phi(\mathfrak{D}) = \underline{o}_{F\times I}(\mathfrak{D})$ is defined by the 1 Czech co-cycle

$$\sigma_{ij} = (F_i/F_j)$$

on $F \times I$. This is computed out as:

$$\begin{aligned}\sigma_{ij} &= (G_i + \varepsilon H_i) \cdot (G_j^{-1}) \cdot (1 - \varepsilon H_j \cdot G_j^{-1}) \\ &= (G_i \cdot G_j^{-1}) \cdot \left[1 + \varepsilon\left(\frac{H_i}{G_i} - \frac{H_j}{G_j} \right)\right] .\end{aligned}$$

Since $\{G_i G_j^{-1}\}$ is a 1-co-cycle defining $\underline{o}_F(D)$, i.e., λ, one translates the I-valued point $\{\sigma_{ij}\}$ back to the origin in $\coprod_\xi P(\xi)$ by dividing by this term. This gives the 1-co-cycle

$$\left[1 + \varepsilon\left(\frac{H_i}{G_i} - \frac{H_j}{G_j} \right)\right]$$

which is the image under the truncated exponential of the 1-co-cycle

$$\tau_{ij} = \left(\frac{H_i}{G_i} - \frac{H_j}{G_j} \right)$$

in $\underline{o}_F$. Then $\{\tau_{ij}\}$ is the point of $H^1(\underline{o}_F)$ corresponding to $\Phi(\overline{\mathfrak{D}})$. On the other hand, $\{\tau_{ij}\}$ is certainly the coboundary of the section $\{H_i/G_i\}$ of N_D.

QED

The final identification is left to the reader to carry out: viz. that if L is an invertible sheaf on F, and if the section $s \in H^0(F, L)$ corresponds to the curve D, hence to the closed point δ in the linear system

$$\mathbf{P} = \mathbf{P}[\widehat{H^0(L)}]$$

of L, then the Zariski tangent space to $\mathbf{P}$ at δ is canonically isomorphic to:

$$H^0(F, L)/k \cdot s \quad .$$

LECTURE 25

THE FUNDAMENTAL THEOREM VIA GROTHENDIECK-CARTIER

Suppose $D \subset F$ is a curve of type ξ, that D corresponds to $\delta \in C(\xi)$, that $\underline{o}_F(D)$ corresponds to $\lambda \in P(\xi)$, and that $L = \underline{o}_F(D)$. <u>If</u> $H^1(F, L) = (0)$, <u>then the following are equivalent:</u>

i) $P(\xi)$ is non-singular at λ,

ii) $C(\xi)$ is non-singular at δ,

iii) $C(\xi)$ is reduced at δ,

iv) $P(\xi)$ is reduced at λ .

<u>Proof</u>: By the results of 1° of the last lecture, there is a neighborhood U of $\lambda \in P(\xi)$ such that the subset $\Phi^{-1}(U)$ of $C(\xi)$ is of the form $\mathbf{P}_N \times U$. This implies that i) and ii) are equivalent, and that iii) and iv) are equivalent. Naturally, i) implies iv). But conversely, since $P(\xi)$ is isomorphic to $P(\tau)$, and $P(\tau)$ is a group scheme, if $P(\xi)$ and hence $P(\tau)$ is reduced, then they are both non-singular (2°, Lecture 24).

In characteristic 0, these conditions <u>always</u> occur because of:

THEOREM 1 (Cartier): Let G be a (algebraic) group scheme over k. If $\operatorname{char}(k) = 0$, then G is non-singular.

<u>Proof</u>: Let υ be the completion of the local ring $\underline{o}_e$ of G at e. Multiplication is a morphism

$$G \times G \xrightarrow{\mu} G$$

such that $\mu(e \times e) = e$: therefore μ defines a homomorphism

$$\mu^*: \upsilon \to [\text{completion of } \underline{o}_{e\times e}] \cong \upsilon \mathop{\hat{\otimes}}_k \upsilon$$

where $\hat{\otimes}$ is the completed tensor product [i.e., use the fact that $\underline{o}_{e\times e}$ is the localization of $\underline{o}_e \otimes \underline{o}_e$ with respect to the maximal ideal $(\underline{o}_e \otimes m_e + m_e \otimes \underline{o}_e)$]. But since μ is a group law, the restriction of μ to either

$$G \times (e) \subset G \times G$$

or

$$(e) \times G \subset G \times G$$

is just the identity from G to G. Algebraically, this means that if you map $\upsilon \hat{\otimes}_k \upsilon$ onto υ by mapping either of the two factors onto its residue field k, and compose this with μ^*, the result is the identity from υ to υ. This means that if $a \in m$, the maximal ideal of υ, then

$$\mu^*(a) - 1 \otimes a - a \otimes 1$$

must go to 0 if either factor of $\upsilon \hat{\otimes}_k \upsilon$ is mapped onto its residue field, i.e.,

$$(\alpha) \qquad \mu^*(a) \in 1 \otimes a + a \otimes 1 + m \hat{\otimes}_k m \quad .$$

We now prove:

(*) for all linear functionals $f\colon\ m/m^2 \to k$, there is a derivation $D\colon\ \upsilon \to \upsilon$ annihilating k and inducing f.

Proof of (*): Extend f to a linear map $F\colon \upsilon \to k$ by requiring that $F = 0$ on k and on m^2. Let D be the composition:

$$D\colon\ \upsilon \xrightarrow{\mu^*} \upsilon \hat{\otimes}_k \upsilon \xrightarrow{1 \otimes F} \upsilon \hat{\otimes}_k k = \upsilon \quad .$$

Then D is clearly linear and D annihilates k. Moreover, by the expression (α), if $a \in m$,

$$\begin{aligned} D(a) &= 1 \otimes F[1 \otimes a + a \otimes 1 + (R)], \qquad R \in m \hat{\otimes} m \\ &= F(a) + (1 \otimes F)(R) \ . \end{aligned}$$

But $(1 \otimes F)(R) \in m$, hence D induces f as a map from m/m^2 to $\upsilon/m = k$. It remains to check:

$$D(a \cdot b) = a \cdot Db + b \cdot Da$$

if $a, b \in m$. But just compute:

$$\begin{aligned} \mu^*(a \cdot b) &= \mu^*(a) \cdot \mu^*(b) \\ &= (1 \otimes a + a \otimes 1 + R) \cdot (1 \otimes b + b \otimes 1 + S) \\ &= (a \otimes 1) \cdot \mu^*(b) + (b \otimes 1) \cdot \mu^*(a) \\ &\qquad + [1 \otimes ab - ab \otimes 1 + R \cdot I \otimes b + S \cdot 1 \otimes a + R \cdot S] \\ &= (a \otimes 1) \cdot \mu^*(b) + (b \otimes 1) \cdot \mu^*(a) + T \end{aligned}$$

where $R, S \in m \hat{\otimes} m$, and $T \in \upsilon \otimes 1 + \upsilon \hat{\otimes} m^2$. Therefore,

$$\begin{aligned} D(a \cdot b) &= (1 \otimes F)[(a \otimes 1) \cdot \mu^* b + (b \otimes 1) \cdot \mu^* a + T] \\ &= a \cdot [(1 \otimes F)\mu^* b] + b \cdot [(1 \otimes F)\mu^* a] \\ &= a \cdot Db + b \cdot Da \ . \end{aligned}$$

To complete the proof of the theorem, let $\overline{X}_1, \ldots, \overline{X}_n$ be a basis of m/m^2. Let $f_1, \ldots, f_n$ be a dual basis, and extend these to derivations $D_1, \ldots, D_n$ of υ. Writing

$$\alpha = (\alpha_1, \ldots, \alpha_n)$$
$$X^\alpha = X_1^{\alpha_1} \cdot \ldots \cdot X_n^{\alpha_n}$$
$$\alpha! = \alpha_1!, \ldots, \alpha_n!$$
$$|\alpha| = \Sigma\, \alpha_i, \quad \alpha_i \geq 0$$
$$D^\alpha f = D_1^{\alpha_1} \cdot \ldots \cdot D_n^{\alpha_n} f$$

we can map υ homomorphically into $k[[X_1, \ldots, X_n]]$ via

$$f \mapsto \sum_{0 \leq |\alpha| < \infty} \frac{\overline{D^\alpha f}}{\alpha!} X^\alpha = A(f)$$

(where $\bar{b}$ is the image of an element $b \in \upsilon$ in k). On the other hand, by the general theory of complete local rings, there is a surjection

$$B\colon \; k[[X_1, \ldots, X_n]] \longrightarrow \upsilon$$

such that $B(X_i) \equiv \bar{x}_i \pmod{m^2}$. Then $A \circ B$ is a homomorphism of $k[[X_1, \ldots, X_n]]$ into itself inducing the identity modulo $(X_1, \ldots, X_n)^2$. Therefore $A \circ B$ is an automorphism; and since B is surjective, this implies that A is an isomorphism.

QED

COROLLARY: If $\mathrm{char}(k) = 0$, then all the schemes $P(\xi)$ are non-singular. Therefore

$$\dim P(\xi) = \dim_k H^1(F, \underline{o}_F) \quad .$$

Proof: By Cartier's theorem, and the isomorphism of the Zariski tangent space of $P(\tau)$ at 0 with $H^1(F, \underline{o}_F)$.

This proves Existence Theorem (A), and re-proves the theorem of Lecture 23, for curves D such that $H^1(F, \underline{o}_F(D)) = (0)$.

170 BL

LECTURE 26

RING SCHEMES; THE WITT SCHEME

§0. Outline

In section 1, the viewpoint of the ring schemes is introduced, with some basic definitions and constructions.

In section 2, we develop the Witt ring scheme associated with a prime p and apply it to the problem for which is was originally used—the inversion of a functor which one would not offhand have suspected was invertible! The problem is developed in parts A and B, the Witt scheme is described in part C, and it is used to solve the problem in part D. The reader wishing to skip this tangential discussion can read part C only.

In section 3, part A, we develop the "universal Witt scheme," a modification of the construction of §2; (a "generalization" in the sense that the Witt scheme associated with any prime p can be gotten by "truncating" the universal scheme). We use it in part B to obtain a "ring of logarithms"—a ring whose additive structure is isomorphic to the multiplicative structure of the set of formal power series (over a given ring R) with first coefficient 1. In parts C, D and E, we describe certain mappings and truncations of the Witt scheme, for which we shall have use later in dealing with power series.

§1. Generalities

In any category C having direct products, and having a final object P, we can define "ring objects": sextuples $(\mathbf{H}, o, \iota, \nu, \alpha, \mu)$, $\mathbf{H}$ an object, o, ι, ν, α, and μ maps:

$$\begin{aligned} o&: P \to \mathbf{H} \text{ (zero element)}\\ \iota&: P \to \mathbf{H} \text{ (unity)}\\ \nu&: \mathbf{H} \to \mathbf{H} \text{ (additive inverse)}\\ \alpha&: \mathbf{H}\times\mathbf{H} \to \mathbf{H} \text{ (addition)}\\ \mu&: \mathbf{H}\times\mathbf{H} \to \mathbf{H} \text{ (multiplication)}\end{aligned}$$

which satisfy the obvious generalizations of the ring axioms for sets and set maps.*

---

* We shall not count $1 \neq 0$ among the ring axioms; we allow the trivial ring.

Given any other object A of our category, we find that a ring structure is induced on $h_H(A)$, so that h_H becomes a contravariant functor from C to Rings.

We are actually already familiar with some examples of ring objects in the category of schemes. The variety of all $n \times n$ matrices is a non-commutative ring scheme. A simpler example is the affine line, which has an obvious ring scheme structure.

Though our definitions hold in the category $\underline{\text{Schemes}}_S$ of schemes over an arbitrary scheme S, we shall here only be working with ring schemes over Spec $\mathbf{Z}$ ("absolute ring schemes") and certain localizations of $\mathbf{Z}$. Also, in all cases which we shall deal with, the underlying schemes will be affine. The maps defining the ring scheme structures will thus be given by ring homomorphisms. These will go in the opposite direction to the scheme maps (since the relation between affine schemes and rings is contravariant) but they will actually be the expected equations, looked at differently. Thus, where we would be accustomed to describing addition on the affine line as the map $(x, x') \to x''$ (sending $\mathbf{A}^1 \times \mathbf{A}^1 \to \mathbf{A}^1$) $x'' = x + x'$, it becomes, in ring terms, the map $\mathbf{Z}[X] \to \mathbf{Z}[X] \otimes \mathbf{Z}[X]$ determined by $X \to X \otimes 1 + 1 \otimes X$.

(A less trivial example is the "Argand plane functor," associating to each ring R the ring of pairs $(x, y) \in R^2$, with termwise addition, and with multiplication given by $(x, y)(x', y') = (xx' - yy', xy' + x'y')$. It is represented by Spec $\mathbf{Z}[X, Y]$ with addition

$$\alpha(X) = X \otimes 1 + 1 \otimes X \quad \text{and multiplication} \quad \mu(X) = X \otimes X - Y \otimes Y$$
$$\alpha(Y) = Y \otimes 1 + 1 \otimes Y \qquad\qquad \mu(Y) = X \otimes Y + Y \otimes X \ .$$

Calling this scheme $\mathfrak{A}$ (for the moment), to what element of the ring $h_{\mathfrak{A}}(\mathfrak{A})$ does the identity map correspond?)

We shall here be interested in ring schemes $\mathbf{H}$ mainly for the sake of the associated functors h_H. The ring schemes represent a certain class of functorial constructions of rings $h_H(R)$ from rings R. (Essentially they give those constructions in which the resulting ring can be described as the set of all n-tuples (n finite or infinite) of members of R satisfying certain polynomial conditions, and where addition and multiplication are given by polynomial functions.)

A ring scheme over some localization of $\mathbf{Z}$ will correspond to a construction in which the polynomials used may involve certain fractional coefficients, and which thus can only be applied to those rings in which certain integers are invertible.*

One functor which it is easy to represent is that associating to a ring R the ring R[[X]] of formal power series in an indeterminate. We shall call the representing ring scheme $\mathbf{V}$. The underlying scheme is Spec $\mathbf{Z}[A_0, A_1, \ldots]$ (where the A's are indeterminates, representing the coefficients of the power series), and the additive and multiplicative maps are given (in terms of the ring $\mathbf{Z}[A_0, A_1, \ldots]$) by

$$\alpha(A_i) = A_i \otimes 1 + 1 \otimes A_i$$

$$\mu(A_i) = \sum_{j=0}^{i} A_j \otimes A_{i-j} \quad .$$

The truncated power series rings, $R[X]/X^n$, are represented by the (finite-dimensional) schemes $V_n = \mathrm{Spec}\ Z[A_0, \ldots, A_{n-1}]$, quotient-ring-schemes of V. These form a projective system: for every pair of positive integers $m \leq n$ there is a truncation map from V_n to V_m corresponding to the inclusion: $Z[A_0, \ldots, A_{m-1}] \subset Z[A_0, \ldots, A_{n-1}]$, and V is the inverse limit of this system.

(Some random notes on representability of functors of Rings → Rings by ring schemes:

Such functors must have the property $h(R \otimes R') = h(R) \otimes h(R')$, hence the functor sending every ring to a fixed ring A cannot be represented. (But one can construct a scheme which sends every ring with connected spectrum to Z —it is a discrete union of copies of Spec Z. If A is infinite, this is not affine, since it is not compact.

If $A \to B$ is a 1-1 map of rings, $h(A) \to h(B)$ must be a 1-1 map of rings. Hence the functor $R \to R/p$ cannot be represented: the 1-1 map $Z \to Q$ gives $Z/p \to 0$.

Though the "power series ring" functor can be represented, the (finite) "polynomial ring" functor apparently can't. What would be a "generic finite polynomial?"!)

§2. P-adic rings and the Witt functor

Most of this material appears in Serre, Corps Locaux, but the presentation there is more rapid, and it is done somewhat differently: the formalism of ring schemes is not there used.

A: Musical Chairs (while shrinking)

Let p be a prime number.

Let A be any ring in which p generates an ideal which is its own radical (i.e., such that A/p has no nilpotents). Then if two elements are in distinct residue classes mod p, so are their p-th powers: $a^p - b^p \equiv (a-b)^p \not\equiv 0 \pmod p$. In other words: the Frobenius endomorphism of A/p is 1-1.

However, if two elements are the same class mod p, their p-th powers will be in the same class mod p^2:

$$(a+px)^p = a^p + (p)a^{p-1}px + \binom{p}{2}a^{p-2}(px)^2 + \ldots \equiv a^p \pmod{p^2} \quad .$$

More generally, replacing $a + px$ by $a + p^k x$ in the above, we see: if two elements are congruent mod $p^k (k \neq 0)$, then their p-th powers will be congruent mod p^{k+1}, whence, by induction, their p^n-th powers will be congruent mod p^{k+n}.

Thus the operation of raising to the p-th power, though it keeps the congruence classes (mod p) distinct, causes each to "shrink down" under the p-adic metric. Since the Frobenius endomorphism of A/p will not, in general, be the identity, these congruence classes will be playing a wild game of musical chairs as they shrink, confusing the situation a bit.

However, suppose that A/p is perfect. (I.e., that the Frobenius endomorphism is 1-1 onto.) Let $[a]$ be any congruence class (mod p) of A. For every n, <u>some</u> congruence class will have its (p^n)-th power in $[a]$. Since the (p^n)-th powers of its members are all congruent to each other mod p^{n+1}, we get a canonical congruence class mod p^{n+1} defined in $[a]$. Furthermore, as n increases, each successive sub-congruence-class in $[a]$ will belong to the preceding.

Clearly, what is being defined is a member of $\hat{A}$, the p-adic completion of A. (We should here assume the p-adic topology separated, to make this meaningful.) Or, assuming A complete to begin with, we get:

LEMMA 1. Let A be a ring complete under the p-adic topology, such that A/p is perfect. Then there is a canonical map $f\colon A/p \to A$ sending each residue class to its unique member which has (p^n)-th roots for all n. f can be characterized as the unique <u>multiplicative</u> homomorphism of $A/p \to A$ which sections the quotient map $A \to A/p$. (Proof of the last sentence left to the reader.)

<u>Example</u>: If A is simply the ring of p-adic integers $f(A/p)$ consists of the $(p-1)$-st roots of unity and zero.

B: <u>The Teichmüller construction</u>

It is well-known that a p-adic number can be represented uniquely by a "power series" $a_0 + a_1p + a_2p^2 + \dots$ where $a_i = 0, 1, \dots, p-1$. But this is of little mathematical interest, because the set of representatives $0, 1, \dots, p-1$ of the residue classes mod p is clearly rather arbitrary.

Now, however, we have a beautiful functorial set of representatives of the residue classes! Making use of them (and generalizing to the rings A dealt with in the previous section — we need only add the hypothesis that p not be a zero divisor in A, so that these power series <u>will</u> be unique), we get:

LEMMA 2: Let A be a complete p-adic ring where p is not a zero-divisor, such that A/p is perfect. Then there is a 1-1 correspondence between members of A and sequences $(\xi_0, \xi_1, \dots)$ of elements of A/p, given by

Suppose we can discover how to calculate in A using these sequence-representations. Then it should follow that we can reconstruct the structure of A from that of A/p!

It will turn out that we can do this, but the results will be in a more convenient form if we use, not precisely the above correspondence, but the correspondence

(1) $$(\xi_0, \xi_1, \xi_2, \ldots) \longleftrightarrow f(\xi_0) + pf(\xi_1^{p^{-1}}) + p^2 f(\xi_2^{p^{-2}}) + \cdots .$$

(This can be seen from the example worked out in Appendix A.)

C: The Witt Scheme (an apparent interlude)

Let $\mathbb{W}$ be the scheme Spec $\mathbb{Z}[X_0, X_1, \ldots]$, and let us map $\mathbb{W}$ into $\mathbb{A}^\infty$ = Spec $\mathbb{Z}[W_0, \ldots]$ by the map given by the Witt polynomials:

(2) $$\begin{aligned} W_0 &= X_0 \\ W_1 &= X_0^p + pX_1 \\ W_2 &= X_0^{p^2} + pX_1^p + p^2 X_2 \\ &\vdots \\ W_n &= X_0^{p^n} + pX_1^{p^{n-1}} + \cdots + p^n X_n \\ &\vdots \end{aligned}$$

(p is still a fixed prime. Note the confusing terminology: the W's are the coordinates of $\mathbb{A}^\infty$, and the X's the coordinates of $\mathbb{W}$.)

Define a ring scheme structure on $\mathbb{A}^\infty$ by the maps

$$\left.\begin{aligned} \alpha(W_s) &= W_s \otimes 1 + 1 \otimes W_s \\ \mu(W_s) &= W_s \otimes W_s \end{aligned}\right\} \text{ all } s.$$

$\mathbb{A}^\infty$ represents the functor that assigns to the ring R, the ring of infinite sequences $(w_0, w_1, \ldots)$ of elements of R under componentwise addition and multiplication, i.e., the direct product of infinitely many copies of R.

We claim that the ring scheme structure on $\mathbb{A}^\infty$ induces a ring scheme structure on $\mathbb{W}$; the unique structure making this map a homomorphism.

To see this, we first note that if we allow ourselves to divide through by p, we can solve the equations (2) for the X's in terms of the W's. This means, in terms of R-valued points, that if p is invertible in R, we can think of the X's and the W's as simply alternative systems of coordinates for elements of the ring R^∞; the W-coordinates are "simpler," in that addition and multiplication in the ring correspond to coordinate-wise addition and multiplication; but we can clearly find polynomial functions to describe these ring operations in terms of the X's—polynomials, we should expect, whose coefficients lie in $\mathbb{Z}[1/p]$.

The "basic miracle" of the Witt rings is that the coefficients of these "arithmetic polynomials" turn out to actually lie in Z. We shall now prove this fact: We let w_n designate the n-th Witt polynomial- $w_n(X_0, \ldots, X_n) = X_0^{p^n} + \ldots + p^n X_n$. For a generalization that will cover all the arithmetic operations (we need addition, multiplication, and—though we skipped mention of it above—additive inverse and the constants 0 and 1), we let Φ designate any polynomial in two variables (one or both of which may, of course, be dummies), with integral coefficients. We know that there will exist polynomials

$$\varphi_0(X_0; X_0'), \ldots, \varphi_n(X_0, \ldots, X_n ; X_0', \ldots, X_n'), \ldots$$

such that for every n,

$$\Phi(w_n(X_0, \ldots, X_n), w_n(X_0', \ldots, X_n')) = w_n(\varphi_0(X_0, X_0'), \ldots, \varphi_n(X_0, \ldots, X_n ; X_0', \ldots, X_n'))$$

or, in abbreviated style, $\Phi(w_n(X), w_n(X')) = w_n(\varphi(X, X'))$. (In using this abbreviated style, when we wish to remind ourselves that $w_n(X)$ involves only $X_0, \ldots, X_n$, we shall write it $w_n(X_{\ldots n})$.)

LEMMA 3: The coefficients of the φ_n are integral.

<u>Sublemma</u>: If $x_i \equiv y_i \pmod{pR}$ $(i = 0, \ldots, n;\ x_i, y_i \in R$, R any ring), then $w_n(x) \equiv w_n(y) \pmod{p^{n+1}R}$.

<u>Proof</u>: Follows immediately from the definition of w_n and the observations of part A of this section.

<u>Proof of the Lemma.</u> Assume the result true for all $i < n$.

We note that $w_n(X) = w_{n-1}(X^p) + p^n X_n$. * Applying this to the right hand side of the equation $\Phi(w_n(X), w_n(X')) = w_n(\varphi(X, X'))$, and solving for $\varphi_n(X, X')$ we get

$$\varphi_n(X, X') = \frac{\Phi(w_n(X), w_n(X')) - w_{n-1}(\varphi^p_{\ldots n-1}(X, X'))}{p^n}$$

This is, in fact, the recursive formula by which the φ's are defined. To show φ_n integral, we must show $\Phi(w_n(X), w_n(X')) \equiv w_{n-1}(\varphi^p(X, X')) \pmod{p^n}$.

---

* If $n = 0$, we interpret w_{-1} as 0. Since this is a polynomial in "those X's with index less than n," and satisfies the sublemma, the proof goes through perfectly well.

Substituting $w_n(X) = w_{n-1}(X^p) + p^n X_n$ in the left-hand side now, and noting that the "$p^n X_n$" terms can be discarded (mod p^n), this side becomes $\Phi(w_{n-1}(X^p), w_{n-1}(X'^p))$. We rewrite this as $w_{n-1}(\varphi(X^p, X'^p))$. To show this congruent to $w_{n-1}(\varphi^p(X, X'))$, it suffices to note (because of the sublemma) that $\varphi_i(X^p, X'^p) \equiv \varphi_i^p(X, X')$ (mod p)—by the Frobenius automorphism. (And that is where we use the inductive assumption that the φ_i are integral.)

QED

The consequence of this is that these polynomials can be used to define a ring structure on the set of infinite sequences of elements in <u>any</u> ring R. The operations will satisfy the ring axioms because they are given by polynomials which satisfy these axioms for all elements of $\mathbf{Z}[1/p]$, and hence satisfy them identically.

The resulting ring will no longer be isomorphic to R^∞; rather, the Witt transformation will give us a <u>homomorphism</u> to R^∞. In the case where p is not a zero-divisor in R, we can see that this transformation is 1-1, so that the Witt ring can be identified with a subring of R^∞. If p is a zero-divisor—e.g., if R is of characteristic p—this too fails to hold.

In scheme-theoretic terms, the above discussion is rendered as follows: We have a map $w\colon \mathbf{W} \to \mathbf{A}^\infty$. Tensoring with $\mathbf{Z}[1/p]$, we discover that

$$w'\colon \mathbf{W} \times \operatorname{Spec} \mathbf{Z}[1/p] \to \mathbf{A}^\infty \times \operatorname{Spec} \mathbf{Z}[1/p]$$

is an isomorphism of schemes (because the system (2) is invertible over $\mathbf{Z}[1/p]$. Hence $\mathbf{A}^\infty \times \operatorname{Spec} \mathbf{Z}[1/p]$'s structure of ring-scheme-over-Spec $\mathbf{Z}[1/p]$ induces a similar structure on $\mathbf{W} \times \operatorname{Spec} \mathbf{Z}[1/p]$. The latter is dense in $\mathbf{W}$, and it turns out that the ring operations extend continuously to all of $\mathbf{W}$. (That is, they are defined by equations with integral coefficients.) These operations will satisfy the ring axioms, because they do so on a dense subset; and since w is a ring homomorphism on a dense subset of $\mathbf{W}$, it is a ring homomorphism. It is clear that the ring scheme structure which we have put on $\mathbf{W}$ is the unique one making w a homomorphism.

The assertions made at the beginning of this section are thus proved.

D: <u>The Grand Finale</u>

We recall the situation of part B.

We claim that the polynomials defining the ring structure of $\mathbf{W}$ are exactly those we need for computing with our "power series." We shall first try to give an intuitive idea why this is so.

A member of the ring A can, we know, be represented in infinitely many ways as a (finite or infinite) power series $a_0 + pa_1 + p^2a_2 + \ldots$. If we think of the unique representation in which each a_i is in $f(A/p)$ as the "correct" representation, and call a representation "correct mod p^n" if each term $p^i a_i$ is congruent mod p^n to the corresponding term of the "correct" expression, then it is easy to check that a sufficient condition for a representation to be correct mod p^n is that each a_i be a (p^{n-i})-th power (for $i < n$).

If we now substitute arbitrary values $x_0, x_1, \ldots$ from A for the $X_0, X_1, \ldots$ of transformation (2);

$$\begin{aligned} w_0 &= x_0 \\ w_1 &= x_0^p + px_1 \\ w_2 &= x_0^{p^2} + px_1^p + p^2x_2 \\ w_3 &= x_0^{p^3} + px_1^{p^2} + p^2x_2^p + p^3x_3 \\ &\vdots \end{aligned}$$

we see that the successive w_i's are more and more nearly "correctly" represented. Looking closely at this situation, we can get some understanding of why the polynomials that tell us what to do with the x's in order to do arithmetic with the w's should also tell us how to handle the terms of the "correct power series representation" of an element, to do arithmetic with that element. The fact that the x's appear with descending exponents matches our use of a representation of the form $f(\xi_0) + pf(\xi_1^{p^{-1}}) + \ldots$ rather than $f(\xi_0) + pf(\xi_1) + \ldots$.

The explicit statement and proof are as follows:

LEMMA 4: Given A and f as in Lemma 1, and Φ and φ_i as in Lemma 3, for all $\xi_0, \xi_1, \ldots; \xi_0', \xi_1', \ldots$ in A/p we have:

$$\Phi\Big(f(\xi_0) + \ldots + p^i f(\xi_i^{p^{-i}}) + \ldots ;\ f(\xi_0') + \ldots + p^i f(\xi_i'^{p^{-i}}) + \ldots\Big)$$

$$= f(\varphi_0(\xi_0;\ \xi_0')) + \ldots + p^i f(\varphi_i(\xi_0, \ldots, \xi_i;\ \xi_0', \ldots, \xi_i')^{p^{-i}}) + \ldots \ .$$

<u>Proof</u>: It will suffice to show that the equation holds mod p^{n+1} for given n. Hence in our calculations, we may discard all terms of the above power series past the "p^n" terms.

Let us substitute $x_i = \xi_i^{p^{-n}}$, $x_i' = \xi_i'^{p^{-n}}$. Noting that p-th power exponents commute with all operations in A/p (by Frobenius), and with f (since it is a multiplicative homomorphism), we rewrite what we are trying to prove as:

$$\Phi(f(x_0)^{p^n} + \ldots + p^i f(x_i)^{p^{n-i}} + \ldots,\ \ldots)$$

$$\equiv f(\varphi_0(x_0, x_0')^{p^n} + \ldots + p^i f(\varphi_i(x_0, \ldots, x_i;\ \ldots))^{p^{n-i}} + \ldots \quad (\text{mod } p^{n+1}) \ .$$

We note that the right-hand side can be rewritten $w_n(f(\varphi(x, x')))$, and the left-hand side as $\Phi(w_n(f(x)), w_n(f(x')))$, which reduces to $w_n(\varphi(f(x), f(x'))$.

By our earlier Sublemma to show these congruent mod p^{n+1}, it suffices to show $f(\varphi_i(x, x')) \equiv \varphi_i(f(x), f(x'))$ (mod p). This is immediate, because f "preserves" congruence class mod p, by construction.

QED

We commented before that if we could find out how to compute with these "power series," we could reconstruct A from A/p. We have now found out how to do this, and we have thus proved:

THEOREM: Let A be as in Lemma 2, $k = A/p$. Then $h_{\mathbf{W}}(k) \cong A$, by a canonical isomorphism.

(It is now not hard to show the converse: that if k is a perfect ring of characteristic p, $h_{\mathbf{W}}(k)$ is a ring where p is not a zero-divisor, complete in the p-adic metric, whose residue ring mod p is k. Needless to say, the functors $h_{\mathbf{W}}$ and "/p" turn out to be (for the rings in question) inverses on the map level as well as the object level. So we get an isomorphism between the category of perfect rings of characteristic p, and a certain category of p-adic rings.)

§3.A. The Universal Witt Scheme

Let us designate the Witt scheme associated with a prime p, described above, $\mathbf{W}^p$; and let us relabel the coordinates $X_0, X_1, \ldots$ and $W_0, W_1, \ldots$ as $X_1, X_p, X_{p^2}, \ldots$ and $W_1, W_p, W_{p^2}, \ldots$. The transformation (2) then becomes:

$$
\begin{aligned}
W_1 &= X_1 \\
W_p &= X_1^p + pX_p \\
W_{p^2} &= X_1^{p^2} + pX_p^p + p^2X_{p^2} \\
&\vdots \\
W_{p^k} &= \sum p^i X_{p^i}^{p^{k-i}} .
\end{aligned}
$$

This family of polynomial functions is clearly a subfamily of one which does not depend on any prime p, namely:

$$\begin{aligned}
W_1 &= X_1 \\
W_2 &= X_1^2 + 2X_2 \\
W_3 &= X_1^3 \qquad\qquad + 3X_3 \\
W_4 &= X_1^4 + 2X_2^2 \qquad\qquad + 4X_4 \\
&\vdots \\
W_n &= \sum_{d|n} dX_d^{n/d} \\
&\vdots
\end{aligned}$$

We shall show for (3) as we did for (2) that the arithmetic operations on the W's correspond to polynomial operations on the X's with integral coefficients.

As before, given Φ, we construct functions φ_n such that $\Phi(w_n(X), w_n(X')) = w_n(\varphi(X, X'))$. Just as our earlier functions "φ_n" depended only on the X_i and X_i' with $i \leq n$, so these φ_n depend only on the X_d and X_d' such that $d|n$. These φ_n could have coefficients in $\mathbf{Q}$, but we find for any prime p:

LEMMA 4': The denominators of the coefficients of φ_n are not divisible by p.

<u>Sublemma</u>: If $p^k|n$, and $x \equiv y \pmod p$, then $w_n(x) \equiv w_n(y) \pmod{p^{k+1}}$.

(Proved as before.)

<u>Proof of the Lemma</u>: Assume the result true for all proper divisors of n.

Let p^k be the greatest power of p dividing n, $n = p^k m$. We note that

$$w_n(X) = w_{n/p}(X^p) + \sum_{d'|m} p^k d'\, X_{p^k d'}^{m/d'} = w_{n/p}(X^p) + p^k(mX_n + \text{terms involving lower } X\text{'s}) .^*$$

Substituting this in the right-hand side of the equation $\Phi(w_n(X), w_n(X')) = w_n(\varphi(X, X'))$, and solving for the last term, we get:

$$m \cdot \varphi_n + \text{terms involving lower } \varphi\text{'s} = \frac{\Phi(w_n(X), w_n(X')) - w_{n/p}(\varphi^p(X, X'))}{p^k}$$

---

* As before, if $p \nmid n$, we set $W_{p/n} = 0$. Note that by "lower X's," we mean, of course, X's whose indices are proper divisors of n.

Since the "lower φ's" are "integral" (i.e., have no denominators divisible by p) by inductive hypothesis, it suffices to show

$$\Phi(w_n(X),\ w_n(X')) \equiv w_{n/p}(\varphi^p(X,\ X')) \pmod{p^k} .$$

This we do exactly as before: we substitute our "$w_n(X) =$" formula in the left-hand side, now discarding the "tail" term since it is divisible by p^k, and "commute" Φ and $w_{n/p}$, so that the desired congruence becomes

$$w_{n/p}(\varphi(X^p,\ X'^p)) \equiv w_{n/p}(\varphi^p(X,\ X')) \pmod{p^k} .$$

This holds by our sublemma. QED

Hence all the coefficients must lie in $\mathbf{Z}$.

So, as before, we get a ring scheme $\mathbf{W}$, with a homomorphism

$$\begin{array}{ccc} \mathbf{W} & \longrightarrow & \mathbf{A}^\infty \\ \| & & \| \\ \operatorname{Spec} \mathbf{Z}[X_1,\ X_2,\ldots] & & \operatorname{Spec} \mathbf{Z}[W_1,\ W_2,\ldots] \end{array}$$

which becomes an isomorphism on tensoring with Spec ($\mathbf{Q}$).

B: Logarithms of power series

Recalling that $\mathbf{V}$ designates the "formal-power-series" ring scheme, let us designate by $\mathbf{V}^\circ$ the closed subscheme corresponding to the equation $A_0 = 1$. This represents power series with constant term 1, and is a commutative group scheme under the restriction of multiplication in $\mathbf{V}$. We shall write the R-valued point $(1,\ a_1,\ a_2,\ldots)$ of $\mathbf{V}^\circ$ in the more familiar form $1 + a_1t + a_2t^2 + \ldots$. We shall deal with $\mathbf{V}^\circ$ in terms of its functor of R-valued points in order to make available to us well-known results about formal power series.

Consider the following maps of schemes:

$$\mathbf{W} \times \operatorname{Spec}(\mathbf{Q}) \xrightarrow[\sim]{w} \mathbf{A}^\infty \times \operatorname{Spec}(\mathbf{Q}) \xrightarrow[\sim]{\psi} \mathbf{V}^\circ \times \operatorname{Spec}(\mathbf{Q})$$

where

$$\psi(w_1,\ w_2,\ldots) = \exp\left[-\sum \frac{w_m}{m}\,t^m\right] .$$

We claim that the composition extends to an isomorphism of the schemes $\mathbf{W}$ and $\mathbf{V}^\circ$. To check this, we first recompute this map on R-valued points, in the case $R \supset \mathbf{Q}$. Say,

$$\sum a_i t^i = \psi(w_1,\ w_2,\ldots) = \psi \circ w(x_1,\ x_2,\ldots) .$$

We get:

$$\begin{aligned}\sum a_i t^i &= \exp\left[-\sum_m \frac{w_m}{m} t^m\right] \\ &= \exp\left[-\sum_m \frac{\sum_{nd=m} n x_n^d}{m} t^m\right] \\ &= \exp\left[-\sum_n \sum_d \frac{(x_n \cdot t^n)^d}{d}\right]. \\ &= \exp\left[\sum_n \log(1 - x_n t^n)\right] \\ &= \prod_{n=1}^{\infty} (1 - x_n t^n) \quad .\end{aligned}$$

The a_i and the x_i are now clearly mutually related by polynomial equations with integral coefficients.

QED

Now the map from $A^\infty \times \operatorname{Spec} Q$ to $V^\circ \times \operatorname{Spec} Q$ is a homomorphism from the additive group structure of the former to the (multiplicative) group structure of the latter. Hence the composite map is such a homomorphism. Hence the scheme-isomorphism between W and V°, being a group homomorphism on a dense subset, is, in fact, an isomorphism of group schemes:

<u>W is a ring scheme whose additive structure is that of the group scheme V°</u>.

(The question "to what operation on V° does the multiplicative structure of W correspond?" is investigated in Appendix B.)

C: <u>Truncations</u>

We can "truncate" the power-series ring-sceme V because its arithmetic operations are such that the n-th term of the sum or product depends only on the n-th and lower terms of the elements given. In W, the n-th term depends only on those terms whose indices <u>divide</u> n. The result is, that given any set S of positive integers which contains every divisor of a number if it contains that number, we get a ring scheme $W_S = \operatorname{Spec} Z[X_s]_{s\in S}$, a "truncation" of W. We shall call such sets S of integers "truncation sets." For any truncation set S, we get a truncation homomorphism $T_S: W \to W_S$.

Various facts are trivial to verify about these schemes: The map $w: W \to A^\infty$ truncates to a map $w_S: W_S \to A^S$, and the ring structure on W_S is the unique structure making this a ring homomorphism. Given two truncation sets $S \subset S'$, we get a truncation homomorphism $T_{S,S'}: W_{S'} \to W_S$, and $T_{S,S'} \circ T_{S',S''} = T_{S,S''}$. W itself is, of course, W_{Z+}, and $W_{\{1,p,p^2,\ldots\}}$ is W^p, the scheme constructed in §2. The scheme $W_{\{1,\ldots,n-1\}}$ are isomorphic to the truncated power series groups V_n°, but the other truncations do not correspond to any familiar construction with power series rings.

We need some general nonsense at this point: A homomorphism $f: A \to B$ of commutative group schemes will be called "1-1" if the induced maps of groups: $h_f(X): h_A(X) \to h_B(X)$ are 1-1 for all schemes X, "onto" if the $h_f(X)$ are all onto.

The property of being 1-1 behaves quite nicely. It is equivalent, by definition, to being a monomorphism in the category of schemes. Given an arbitrary homomorphism $f: A \to B$, we can get a 1-1 homomorphism $K \to A$ whose functor is the functor of kernels of the induced group maps. (We construct K as the fibre in A of the Z-valued point 0 of B. How do we show that the group operation lifts to K ?)

On the other hand, the property we have called being "onto" is stronger than being an epimorphism both of schemes and of group schemes. It is equivalent to the existence of a scheme map g from B back to A which "sections" f — a right inverse map. This is clearly sufficient; to see that it is necessary, we note that by our definition of "onto", the identity map in $h_B(B)$ must come from a map g in $h_A(B)$ such that $fg =$ identity. (But g will not in general be a group scheme homomorphism!)

We cannot in general construct a group scheme with the properties of a cokernel of f. Hence, though exact sequences can be <u>defined</u> (by the condition that the induced sequences $\to h_A(X) \to h_B(X) \to h_C(X) \to$ all be exact—note that this implies that the kernel of each map <u>is</u> a cokernel to the preceding), they are not so easy to come by. However, <u>given</u> an onto map $A \to B \to 0$, we can get an exact sequence $0 \to \operatorname{Ker} f \to A \to B \to 0$.

Note that the conditions "1-1," "onto" and "exact" respect base extension.

The truncation maps we have defined are all onto: Given $S \subset S'$, we get a section W_S back to $W_{S'}$ by "filling in" the missing coordinates $X_{s'}$ in any way we like, e.g., with zeroes.

D: <u>Canonical maps</u>

There are two sets of maps from W to W which are useful.

a) Define V_n: $\mathbf{W} \to \mathbf{W}$ by

$$V_n^*(X_m) = \begin{cases} X_{m/n} & \text{if } n|m \\ 0 & \text{otherwise.} \end{cases}$$

(In terms of R-valued points, for instance, $V_3(x_1, x_2, \ldots) = (0, 0, x_1, 0, 0, x_2, \ldots)$.) We claim:

i) $V_n \circ V_m = V_{nm}$.

ii) V_n is an additive isomorphism from $\mathbf{W}$ onto the kernel of the truncation $T_{Z^+-n\,Z^+}$

i) is obvious, and one checks that V_n is at least an isomorphism of the scheme $\mathbf{W}$ with this kernel by looking at R-valued points. To check the additiveness, it suffices to tensor with $\mathbf{Q}$ and show that the induced map on $\mathbf{A}^\infty \times \operatorname{Spec} \mathbf{Q}$ is additive. We find, in fact, that it is described by

$$W_m \mapsto \begin{cases} nW_{m/n} & \text{if } n|m \\ 0 & \text{otherwise.} \end{cases}$$

QED

For any truncation set S, we observe that we have, similarly, a map

$$V_{S,n}\colon \mathbf{W}_{S/n} \to \mathbf{W}_S$$

where $S/n = \{m \in Z | nm \in S\}$, which identifies $\mathbf{W}_{S/n}$ with the kernel of the trucation

$$\mathbf{W}_S \to \mathbf{W}_{S-nZ^+} \quad .$$

b) Define F_n: $\mathbf{W} \to \mathbf{W}$ by its action on R-valued points of the isomorphic scheme $\mathbf{V}^\circ$: let P(t) be a power series in t with first coefficient 1. Let us designate by $\tau_1, \ldots, \tau_n$ the formal n-th roots of t; then the product

$$\prod_i P(\tau_i)$$

being symmetric in the τ's, will again be a power series in t, and its coefficients will be polynomials in the coefficients of P. An examination of the map defining the relation between $\mathbf{V}^\circ$ and $\mathbf{A}^\infty$ shows us that F_n corresponds to the map

$$(w_1, w_2, \ldots) \to (w_n, w_{2n}, \ldots)$$

of R-valued points of $\mathbf{A}^\infty$. We note that this is a ring homomorphism, so F_n is a ring homomorphism. Also $F_n \circ F_m = F_{nm}$.

We deduce (by the usual "only-those-indices-that-devide-m" arguments) that similar maps (also ring homomorphism) are defined between the truncated schemes:

$$F_{S,n}: \mathbf{W}_S \to \mathbf{W}_{S/n} \quad .$$

c) Look at $F_n \circ V_n$: checking it on R-valued points of $\mathbf{A}^\infty$, we find:

$$F_n \circ V_n = \text{multiplication by } n.^*$$

In some cases, one can divide by n:

LEMMA 5: n is invertible in $\mathbf{W} \times \operatorname{Spec} \mathbf{Z}[1/n]$.

Proof: We recall that one can take n-th roots of monic power series if we allow division of the coefficients by n; hence one can divide by n in $\mathbf{W} \times \operatorname{Spec} \mathbf{Z}[1/n]$.

QED

Thus, over $\operatorname{Spec} \mathbf{Z}[1/n]$, $\frac{1}{n} V_n$ is a right inverse to F_n.

E: Direct product decompositions

The direct product of two ring schemes $\mathbf{H}$ and $\mathbf{H}'$ over S is defined just like the direct product of two rings. (Do not confuse this with the tensor product!) Its underlying scheme is the product over S of the schemes for $\mathbf{H}$ and $\mathbf{H}'$.

Starting with a commutative ring scheme $\mathbf{G}$, there is a 1-1 correspondence between decompositions $\mathbf{G} = \mathbf{H} \times \mathbf{H}'$ and S-valued idempotent points ε in $\mathbf{G}$: the element $(1, 0)$ of $\mathbf{H} \times \mathbf{H}'$ is an ε, and $\mathbf{H}$ and $\mathbf{H}'$ are the kernels of multiplication by $1-\varepsilon$ and ε respectively.

Look at $\mathbf{A}^\infty = \operatorname{Spec} \mathbf{Z}[W_1, W_2, \ldots]$ over $\operatorname{Spec} \mathbf{Z}$. For every subset I of the positive integers, $\mathbf{A}^\infty$ has a ($\mathbf{Z}$-valued) idempotent point η_I:

$$\eta_I^*(W_i) = 1 \qquad i \in I$$
$$= 0 \qquad i \notin I$$

and correspondingly decomposes:

$$\mathbf{A}^\infty = \mathbf{A}^I \times \mathbf{A}^{\mathbf{Z}^+ - I} \quad .$$

---

* We mean, of course, the ring-scheme operation of multiplication by n, which does not correspond to coordinate-wise multiplication by n except for the coordinates w_1 of $\mathbf{A}^\infty$. The same should be understood in the following lemma, concerning multiplication by the $\operatorname{Spec} \mathbf{Z}[1/n]$-valued point "$1/n$."

Hence W admits all these decompositions too, <u>over</u> Spec Q. The question arises: suppose P is any set of primes, and

$$\mathcal{P} = \operatorname{Spec} Z[\ldots, 1/p, \ldots]_{p \notin P} .$$

Then how many of these decompositions of $W \otimes \operatorname{Spec} Q$ actually occur over $\mathcal{P}$? Equivalently, which of the $\varepsilon_I = w^{-1}(\eta_I)$ are rational over $\mathcal{P}$ — have no primes in P occurring in the denominators of their coordinates?

Clearly, if we replace W by a truncation W_S the same questions can be asked for subsets $I \subset S$.

Let Q be the set of primes not in P. Let $\overline{P}$ (respectively $\overline{Q}$) designate the multiplicative semigroup of positive integers generated by P (respectively Q) and 1. Note that the sets $n\overline{P}$ for $n \in \overline{Q}$ partition Z^+ .

LEMMA 6: For any $n \in \overline{Q}$, $\varepsilon_{n\overline{P}} \in W$ is rational over $\mathcal{P}$.

<u>Proof</u>: For any $n \in \overline{Q}$, we note that the projection given by ε_{nZ^+} is simply $\frac{1}{n} V_n F_n$, hence is rational over $\mathcal{P}$; in particular, ε_{nZ^+} itself is rational. Now

$$\varepsilon_{n\overline{P}} = \prod_{p \in Q} (\varepsilon_{nZ^+} - \varepsilon_{pnZ^+}) .$$

This is, formally, an infinite product. However, it "converges" coordinatewise in the sense that each coordinate is constant after a certain number of terms. This is clear in the A^∞ coordinates, hence it is also true in the W-coordinates. Hence the left-hand term is rational.

COROLLARY: For any truncation set S and $n \in \overline{Q}$, $\varepsilon_{n\overline{P} \cap S} \in W_S$ is rational over $\mathcal{P}$.

LEMMA 7: Let S be a truncation set. Then over $\mathcal{P}$

$$W_S = \mathop{\mathsf{X}}_{n \in \overline{Q}} \varepsilon_{n\overline{P} \cap S}(W_S) \quad \text{(all schemes tensored with } \mathcal{P}).$$

<u>Proof</u>: If our set of idempotents were finite, the method of proof would be clear. It turns out that we can here apply the same proof without finiteness. We are to verify the universal property of products on X-valued points. Given a family of maps $\alpha_n \colon X \to \varepsilon_{n\overline{P} \cap S}(W_S)$, we take the map $\Sigma_{\overline{Q}}\, \alpha_n \colon X \to W_S$. This infinite sum is defined by exactly the same reasoning used in the last lemma, and is clearly the unique map whose compositions with the projections give the α_n.

COROLLARY: If a set I is the union of sets $n\bar{P} \cap S$ for certain $n \in \bar{Q}$, then ε_I is rational over $\mathcal{P}$.

LEMMA 8: Let $n \in \bar{Q}$, and S be any truncation set. Then

$$\varepsilon_{n\bar{P}\cap S}(\mathbf{W}_S) \cong \mathbf{W}_{\bar{P}\cap S/n} \quad \text{(all schemes tensored with } \mathcal{P}\text{)}.$$

<u>Proof</u>: Consider the maps

$$\varepsilon_{n\bar{P}\cap S}(\mathbf{W}_S) \; \underset{\text{projection}}{\overset{\text{inclusion}}{\rightleftarrows}} \; \mathbf{W}_S \; \underset{(1/n)V_{S,n}}{\overset{F_{S,n}}{\rightleftarrows}} \; \mathbf{W}_{S/n} \; \underset{\text{any section of truncation}}{\overset{\text{truncation}}{\rightleftarrows}} \; \mathbf{W}_{\bar{P}\cap S/n}$$

All are rational over $\mathcal{P}$. It will suffice to show that the composition of the maps going to the right and the composition of the maps going to the left are ring scheme homomorphisms, and are inverses to one another. Tensoring with Spec $\mathbf{Q}$ and using the A-coordinates, we verify easily that this is so.

QED

Hence we have, for every truncation set S and set of primes, P:

$$\mathbf{W}_S \otimes \mathcal{P} = \underset{n\in\bar{Q}}{\mathsf{X}} \varepsilon_{n\bar{P}\cap S}(\mathbf{W}_S \otimes \mathcal{P}) \cong \underset{n\in\bar{Q}}{\mathsf{X}} \mathbf{W}_{\bar{P}\cap S/n} \otimes \mathcal{P}$$

(One might want to know whether what we have achieved is always a maximal direct product decomposition of $W_S \otimes \mathcal{P}$; equivalently, whether the ε_I, for I a union of sets $n\bar{P} \cap S$ $(n \in \bar{Q})$ are the only idempotents of $\mathbf{W}_S$. We prove in Appendix C that this is so.)

A). (Cf. end of §2B, p. 175)

We shall figure out explicitly how to add the first two terms of series of the type originally proposed $((1_0))$. What we must do is solve, for s_0 and s_1, the congruence:

$$(f(a) + pf(b)) + (f(a') + pf(b')) \equiv f(s_0) + pf(s_1) \pmod{p^2}.$$

Reducing mod p, and recalling that $f(a)$ belongs to the residue class a, we get $a + a' = s_0$.

Substituting this back in, and isolating the term in s_1, we get

$$pf(s_1) \equiv pf(b) + pf(b') + (f(a) + f(a') - f(a+a')) \pmod{p^2}.$$

We know that the last expression is a multiple of p. If we could express it as such, we could "divide through" by p and would be finished. The problem is to get an expression for $f(a+a')$. The solution is as follows: $(f(a^{1/p}) + f(a'^{1/p}))^p$ belongs to the congruence class $a + a'$, and, being a p-th power, must belong to the subclass mod p^2 of $f(a+a')$!

Now $(x+y)^p$ can be written $x^p + y^p + p[x, y]$, where $[x, y]$ is a polynomial in x and y with integral coefficients. Hence $f(a+a') \equiv (f(a^{1/p}) + f(a'^{1/p}))^p \equiv f(a) + f(a') + p[f(a^{1/p}), f(a'^{1/p})]$. Hence

$$pf(s_1) \equiv pf(b) + pf(b') - p[f(a^{1/p}), f(a'^{1/p})] \pmod{p^2}$$

$$f(s_1) \equiv f(b) + f(b') - [f(a^{1/p}), f(a'^{1/p})] \pmod{p}$$

$$s_1 = b + b' - [a^{1/p}, a'^{1/p}] .$$

So $(a, b, \ldots) + (a', b', \ldots) = (a+a', b+b' - [a^{1/p}, a'^{1/p}], \ldots)$.

If we would like a set of coordinates in which we can compute purely by polynomial operations, we should either substitute $a = \alpha^p$ or substitute $b = \beta^{1/p}$. The first choice would be unwise, since when we bring in the third term of the expansion, we would have to change again, and so on. The second choice is the one we made in the text. In terms of the expansion (1), the above result is:

$$(\alpha, \beta, \ldots) + (\alpha', \beta', \ldots) = (\alpha+\alpha', \beta+\beta' - [\alpha, \alpha'], \ldots) .$$

B). (Cf. end of §3b, p. 182)

We want to investigate the "multiplication" induced on $V°$ by the isomorphism with W. We shall, as usual, look at R-valued points. What we have to describe is then a binary operation on power series, which we shall write "$\circ$".

We find first of all, using the formula for the isomorphism between $A^{\infty} \times \operatorname{Spec} Q$ and $V° \times \operatorname{Spec} Q$ that $(1-at)\circ(1-bt) = 1 - (ab)t$, when a and b are members of any ring containing Q. It follows that this must hold for a and b in any ring. Since $\circ$ distributes with respect to multiplication, we get

$$\prod^{m} (1 - \alpha_i t) \circ \prod^{n} (1 - \beta_j t) = \prod^{m,n} (1 - \alpha_i \beta_j t) .$$

For the sake of simplicity, let us call the α_i (rather than the $1/\alpha_i$) the "roots" of $\Pi^m (1 - \alpha_i t)$. (Under this definition, a polynomial has an indefinite number of zero roots.) Over an algebraically closed field k, then, we can describe $\circ$ precisely for the finite (i.e., terminating) power series: it is the function sending any pair of polynomials to the polynomial whose roots are all the pairwise products of those of the given two. It is easy to see from this that the rational functions (quotients of polynomials) form a subring, which has, in fact, the structure of the "group ring" on the group k.*

The full power series ring is the completion of this ring under a metric that makes two points "close" if the first n symmetric functions on them agree (though, of course, it takes some rigging to define the "symmetric functions" on a family some of whose members occur with negative multiplicity).

This interpretation goes through in a more or less formal way for any ring without zero divisors. We can construct over any such ring a unique polynomial whose roots are all the pairwise products of the roots of two given polynomials, even if these roots don't lie in the ring itself. The rational functions in the monic-formal-power-series group form a subring which can be thought of as the "semigroup ring" on the nonzero elements—

* It is amusing to note that a somewhat similar construction turns up in algebraic topology. A complex vector bundle on a space X induces a "Chern class" polynomial over the ring $H^{even}(X)$. It turns out that the operation "$\oplus$" on bundles corresponds to multiplication of polynomials, while the taking of tensor products of bundles corresponds to the operation associating to a pair of polynomials the polynomial whose roots are all the pairwise sums of the roots ("in" H^2) of the given two! Such an operation cannot be defined in our power-series context, because the "indefinite number of zero roots," which can be ignored under our "multiplicative multiplication," wreaks havoc with an attempt to set up an "additive multiplication." The essence of the problem is that our polynomials are of indefinite degree in t, while the topologist's polynomials have a definite degree, corresponding to the dimension of the bundle.

but we now must allow not only formal sums of elements actually in the ring, but also sums of elements (integrally) algebraic over the ring, so long as they appear in full sets of conjugates. The full ring is again a completion.

The w_n — the coordinates of the image in $\mathbb{A}^{\infty}$ — are the moments $\Sigma\, \alpha_i^n$.

C). We shall sketch a proof that the direct product decomposition of $\mathbb{W}_S \otimes \mathcal{P}$ given in our final theorem is maximal.

We first note that every idempotent of $\mathbb{W}_S$ over $\mathcal{P}$ gives an idempotent of $\mathbb{A}^S$, and the only idempotents of the latter are η_I for subsets I of S; hence the only possible idempotents in the former are the ε_I. What we desire to show then is that ε_I is rational over $\mathcal{P}$ if and only if I is a union of sets $n\overline{P} \cap S$ $(n \in \overline{Q})$. An equivalent statement is: for every $p \in P$ and elements m, $pm \in S$, we have $m \in I \Longleftrightarrow pm \in I$.

It clearly suffices to check this in the case $P = \{p\}$, a singleton. So suppose we had a rational ε_I with I not satisfying this condition. Then there would exist $m \in \overline{Q}$ and k greater than zero such that $m, pm, \ldots, p^{k-1}m \in I$, $p^k m \in S - I$ (interchanging I and $S-I$ if necessary). Consider the factor of $\mathbb{W}_S$ (we shall drop the "$\otimes\, \mathcal{P}$'s" for convenience) corresponding to $m\overline{P} \cap S$. It will be isomorphic to $W_{\overline{P} \cap S/m}$, a truncation of which is $W_{\{1,p,\ldots,p^k\}}$. If we now follow our idempotent ε_I through all these transformations, we find that it gives us a direct product decomposition of this scheme from which it can be deduced that the truncation:

$$\mathbb{W}_{\{1,\ldots,p^k\}} \rightarrow \mathbb{W}_{\{1,\ldots,p^{k-1}\}} \quad \text{(all schemes tensored with } \mathcal{P}\text{)}$$

splits. But if we take $\mathbb{Z}/p$-valued points, this means by the results of §2D that:

$$\mathbb{Z}/p^k \rightarrow \mathbb{Z}/p^{k-1} \quad \text{splits.} \qquad \text{Contradiction!}$$

LECTURE 27

THE FUNDAMENTAL THEOREM IN CHARACTERISTIC p

1°. Let $\mathbf{H}$ be any ring scheme over the field k. Then, for all schemes X over k, $\mathbf{H}$ defines a sheaf of rings $< \mathbf{H} >_X$ on X via

$$\Gamma(U, < \mathbf{H} >_X) = \mathrm{Hom}_k(U, \mathbf{H}) .$$

In particular, if $\mathbf{A}^1$ is given its canonical ring scheme structure, then

$$< \mathbf{A}^1 >_X \cong \underline{o}_X ,$$

i.e., we recover the structure sheaf on X. On the other hand, suppose the characteristic is p. Then using the Witt ring-scheme for p, we can get an interesting sheaf of rings,

$$\mathfrak{W}_{\infty,X} = < \mathbf{W}_{\{1,p,p^2,\ldots\}} \times \mathrm{Spec}\ k >_X .$$

Similarly, for every finite n, we get a sheaf of rings from the truncated scheme:

$$\mathfrak{W}_{n,X} = < \mathbf{W}_{\{1,p,p^2,\ldots,p^{n-1}\}} \times \mathrm{Spec}\ k >_X .$$

These sheaves of rings form a projective system of sheaves, under the obvious truncations

$$T_{n,n'}: \mathfrak{W}_{n,X} \to \mathfrak{W}_{n',X} \qquad (n > n') ,$$

with inverse limit $\mathfrak{W}_{\infty,X}$, and with first term $\mathfrak{W}_{1,X} = < \mathbf{W}_{\{1\}} >_X = < \mathbf{A}^1 >_X = \underline{o}_X$. These sheaves were introduced by Serre at the Mexico Conference in Topology (1956). To describe their cohomology, Serre introduced certain fundamental homomorphisms called the <u>Bockstein operations</u>. To understand these, it is convenient to take a very general functorial setting:

Say $\mathcal{C}, \mathcal{C}'$ are two abelian categories, and
$F: \mathcal{C} \to \mathcal{C}'$ is a left exact functor with derived functors R^iF.

Assume that a) $\{A_n\}_{n \in \mathbf{Z}^+}$ and

b) surjective homomorphisms $A_n \to A_{n'}$, all $n' \leq n$ form an inverse system.

Let $A_0 = (0)$, and $A_n \to A_0$ be the 0 homomorphism. Let $K_{n,n'}$ be the kernel of $A_n \to A_{n'}$. Then there is a spectral sequence, with

$$E_1^{p,q} = R^{p+q} F(K_{p+1,p}) .$$

(Warning: this p is not the characteristic of k.)

In fact, if

$$B_r^{p,q} = \text{Ker } R^{p+q} F(K_{p+1,p}) \to R^{p+q} F(K_{p+1,p-r+1})$$

$$Z_r^{p,q} = \text{Im } R^{p+q} F(K_{p+r,p}) \to R^{p+q} F(K_{p+1,p})$$

(with respect to the obvious maps) then one checks that

$$(0) = B_1^{p,q} \subset B_2^{p,q} \subset B_3^{p,q} \subset \ldots \subset Z_3^{p,q} \subset Z_2^{p,q} \subset Z_1^{p,q} = E_1^{p,q} .$$

Then, by definition,

$$E_r^{p,q} = Z_r^{p,q}/B_r^{p,q} .$$

Moreover, one finds that there are canonical homomorphisms

$$d_r\colon\ Z_r^{p,q} \to E_1^{p+r,q-r+1} / B_r^{p+r,q-r+1} .$$

Its kernel is the next Z, viz. $Z_{r+1}^{p,q}$, and its image is the next B, viz.:

$$B_{r+1}^{p+r,q-r+1} / B_r^{p+r,q-r+1} .$$

I.e., each successive d is defined on the kernal of the previous d, with values modulo the image of the previous d. This is exactly a spectral sequence.

[For details, the best source appears in the Séminaire Cartan, 1950/51, exposé 8; however, as a matter of my own experience, it is easier and more helpful to work these things out oneself for small r, rather than to follow someone else's sub and super-scripts in detail.]

I leave it to the reader to check that, in good cases, the sequence abuts to

$$\varprojlim_{p} R^n F(A_p) .$$

We want to apply this machine to give a criterion for an element of $H^1(X, \underline{o}_X)$ to lift to $H^1(X, \mathfrak{O}_{\infty,X})$: i.e., take $\mathcal{C}$ as the category of sheaves of abelian groups on X, $\mathcal{C}'$ as the category of abelian groups, F as $H^0(X, \cdot)$, and A_n as $\mathfrak{O}_{n,X}$. Then

$$E_1^{p,q} = H^{p+q}(X, \text{Ker } (\mathfrak{O}_{p+1,X} \to \mathfrak{O}_{p,X})) .$$

In particular,

$$E_1^{o,q} = H^q(X, \mathfrak{W}_{1,X}) = H^q(X, \underline{o}_X)$$

and $Z_r^{o,q}$ is the subgroup of $H^q(X, \underline{o}_X)$ which lifts to $H^q(X, \mathfrak{W}_{r,X})$.

Definition: The homomorphisms d_r on $Z_r^{o,q} \subset H^q(X, \underline{o}_X)$ are called the Bockstein operations β_r.

The point is:

(*) $$\bigcap_r \ker(\beta_r) = \left\{ x \in H^q(X, \underline{o}_X) \,\middle|\, \begin{matrix} x \text{ lifts to } H^q(X, \mathfrak{W}_{r,X}) \\ \text{for all } r \end{matrix} \right\}$$

To have a better understanding of this apparatus, we need one more fact:

LEMMA: $\mathrm{Ker}\,\{\mathfrak{W}_{n+1,X} \to \mathfrak{W}_{n,X}\} \cong \underline{o}_X$.

Proof: This follows immediately from the corresponding result on Witt ring schemes, viz., the kernel of the truncation:

$$W_{\{1,p,p^2,\ldots,p^n\}} \to W_{\{1,p,p^2,\ldots,p^{n-1}\}}$$

is isomorphic, as additive group scheme, to A^1. This was remarked in Lecture 26, §3D (a) (take V_{p^n}).

QED

Therefore, β_{r+1} is a canonical homomorphism:

$$\begin{matrix} \mathrm{Ker}(\beta_r) & \to & H^{q+1}(X, \underline{o}_X)/\mathrm{Im}(\beta_r) \\ \cap & & \\ H^q(X, \underline{o}_X) & & \end{matrix}$$

2°. Let F be a non-singular projective surface over k (actually neither the non-singularity, nor the dimension being 2 is essential). We can now prove the fundamental theorem concerning the families of curves on F when $\mathrm{char}(k) = p$. Let P be the connected component of the identity in the Picard scheme of F. We know from Lecture 24 that the tangent space $T_{o,P}$ to P at o is canonically isomorphic to $H^1(F, \underline{o}_F)$: via this identification—

THEOREM: The tangent space to P_{red} corresponds to the subspace of $H^1(F, \underline{o}_F)$ annihilated by all the Bockstein operations.

<u>Proof</u>: Let $t \in T_{o,P}$. Let

$$I_{(n)} = \text{Spec } k[\varepsilon]/(\varepsilon^n) \ ,$$

and let t correspond to the homomorphism

$$h_2: \ I_{(2)} \longrightarrow P$$

a) t is tangent to P_{red} if and only if, for all n, h_2 lifts to a morphism h_n:

$$\begin{array}{ccc} I_{(2)} & \xrightarrow{\ h_2\ } & P \\ \cap & \nearrow_{h_n} & \\ I_{(n)} & & \end{array} \quad .$$

<u>Proof</u> <u>of</u> a): In terms of local rings, let $\upsilon = \underline{o}_{0,P}$, and let h_2 and t define

$$f_2: \ \upsilon \longrightarrow k[\varepsilon]/\varepsilon^2 \quad .$$

Let $\mathfrak{N} \subset \upsilon$ be the ideal of nilpotnets. Then if t is tangent to P_{red}, it follows that $f_2(\mathfrak{N}) = 0$. Since $\upsilon/\mathfrak{N}$ is regular, by the Proposition in (A), Lecture 22, f_2 lifts to f_n:

$$\begin{array}{ccccc} \upsilon & \longrightarrow & \upsilon/\mathfrak{N} & \xrightarrow{\ f_2\ } & k[\varepsilon]/(\varepsilon^2) \\ & & & \searrow^{f_n} & \uparrow \\ & & & & k[\varepsilon]/(\varepsilon^n) \end{array}$$

hence h_2 lifts to h_n. Conversely, if h_2 lifts to h_n, then f_2 lifts, for every n, to an f_n. Suppose $x \in \mathfrak{N}$; then $x^m = 0$ for some m. Let $f_2(x) = \alpha \cdot \varepsilon$, $\alpha \in k$. Then

$$\begin{aligned} 0 = f_{m+1}(x^m) &= [f_{m+1}(x)]^m \\ &= [\alpha \cdot \varepsilon + \dots]^m \\ &= \alpha^m \cdot \varepsilon^m \quad . \end{aligned}$$

Therefore $\alpha^m = 0$, hence $\alpha = 0$. This means that f_2 annihilates $\mathfrak{N}$, i.e., t is tangent to P_{red}.

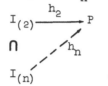

Now translate this into functors: for all n,

$$\begin{array}{c} \mathrm{Hom}(I_{(n)}, P) \subset \mathrm{Hom}(I_{(n)}, \coprod_{\xi} P(\xi)) \\ \wr\| \\ \underline{\mathrm{Pic}}_F(I_{(n)}) \\ \wr\| \\ H^1(F, \underline{o}^*_{F \times I_{(n)}}) \\ \wr\| \\ H^1(F, (\underline{o}_F \otimes k[\varepsilon]/\varepsilon^n)^*) \quad . \end{array}$$

But $[\underline{o}_F \otimes k[\varepsilon]/\varepsilon^n]^* \cong \underline{o}^*_F \cdot [1 + \underline{o}_F \otimes (\varepsilon)/(\varepsilon^n)]$ where (ε) denotes the ideal generated by ε. Therefore

$$H^1(F, [\underline{o}_F \otimes k[\varepsilon]/\varepsilon^n]^*) \cong H^1(F, \underline{o}^*_F) \oplus H^1(F, 1+\underline{o}_F \otimes \frac{(\varepsilon)}{(\varepsilon^n)}) \quad .$$

It follows that:

$$\begin{array}{c} \text{Subgroup of } I_{(n)}\text{-valued points of } P \text{ at } 0 \\ \wr\| \\ H^1(F, 1 + \underline{o}_F \otimes \frac{(\varepsilon)}{(\varepsilon^n)}) \\ \wr\| \\ H^1(F, < V_n^o >_F) \\ \wr\| \\ H^1(F, < \mathbb{W}_{\{1,2,\dots,n-1\}} >_F) \quad . \end{array}$$

Now we use the results of Lecture 26, (E). We are working with the Witt ring scheme over a field of characteristic p, so every prime except p is invertible. Therefore $\mathbb{W}$ decomposes as in (E), with

$P = \{p\}$ $\qquad\qquad \overline{P} = \{1, p, p^2, \dots,\}$

$Q =$ all primes but p; $\qquad \overline{Q} =$ integers prime to p .

Therefore, if $p^\ell \leq n-1$ and $p^{\ell+1} \geq n$, we get:

b) Via the truncation:

$$\mathbb{W}_{\{1,2,\dots,n-1\}} \times \mathrm{Spec}(k) \to \mathbb{W}_{\{1,p,p^2,\dots,p^\ell\}} \times \mathrm{Spec}(k)$$

the latter ring scheme is a <u>direct summand</u> of the former.

Therefore, for every n, we get a diagram:

$$
\begin{array}{ccccc}
\left\{\begin{array}{c} I_{(n)}\text{-valued} \\ \text{points of } P \\ \text{at } 0 \end{array}\right\} & \cong & H^1(F, \langle W_{\{1,2,\ldots,n-1\}}\rangle_F) & & \\
\Big\downarrow \text{res} & & \searrow \;\nwarrow\!\text{-}\,\text{-} & & \\
 & & H^1(F, \langle W_{\{1,p,\ldots,p^{\ell}\}}\rangle_F) & \cong & H^1(F, \mathcal{W}_{\ell,F}) \\
 & & \swarrow & & \Big\downarrow \\
\left\{\begin{array}{c} I_{(2)}\text{-valued} \\ \text{points of } P \\ \text{at } 0 \end{array}\right\} & \cong & H^1(F, \langle W_{\{1\}}\rangle_F) & \cong & H^1(F, \underline{o}_F)
\end{array}
$$

This shows that an element $\alpha \in H^1(F, \underline{o}_F)$ lifts to $H^1(F, \mathcal{W}_{\ell,F})$ (for all ℓ) if and only if it lifts to $H^1(F, \langle W_{\{1,2,\ldots,n-1\}}\rangle_F)$ (for all n); and that this occurs if and only if the corresponding tangent vector t to P at 0 lifts to an $I_{(n)}$-valued point of P (for all n). By a), the theorem is proven.

QED

COROLLARY: P is reduced if and only if all the Bockstein operations from $H^1(F, \underline{o}_F)$ to $H^2(F, \underline{o}_F)$ are 0.

COROLLARY: Let $D \subset F$ be a curve such that $H^1(F, \underline{o}_F(D)) = (0)$. Let $\delta \in C(\xi)$ be the corresponding point. Then $C(\xi)$ is reduced if and only if the same Bockstein operations are 0.

COROLLARY (Severi-Nakai): If $H^2(F, \underline{o}_F) = (0)$, then P is reduced, and the same existence theorems as in char(0) are valid.

For examples where the Bockstein operations are non-zero, see:

Igusa, Proc. Natl. Acad. Sciences, 1953.
Serre, International Symposium in Algebraic Topology, Mexico, 1956.

WORKS REFERRED TO

[1] A. Andreotti and P. Salmon, Anelli con unica decomponibilità in fattori primi ed un problema di intersezioni complete, Monatshefte für Math., 1957 (61), p. 97.

[2] N. Bourbaki, "Algèbra commutative," fasc. 27, 28 and 30 of <u>Èléments de Mathématique</u>, Hermann, Paris, 1961-64.

[3] E. Brown, Cohomology Theories, Annals of Math., 1962 (75), p. 467.

[4] H. Cartan, "Séminaire," Mimeographed notes, obtainable occasionally from the Sécretariat mathématique, Paris.

[5] R. Godement, <u>Théorie des faisceaux</u>, Hermann, Paris, 1958.

[6] EGA: A. Grothendieck, "Élements de géométrie algébrique," Publ. Math. de l'Inst. des Hautes Ét. Sci., Paris, No. 4, 8, 11, 17, 20, 24 etc.

[7] SGA: A. Grothendieck, "Séminaire de géométrie algébrique," Inst. des Hautes Ét. Sci., Paris, 1960-61.

[8] A. Grothendieck, "Fondements de la géométrie algébrique," mimeographed notes sometimes obtainable from the Secrétariat mathématique, Paris.

[9] TOHOKU: A. Grothendieck, "Sur quelques points d'algèbra homologique," Tôhoku Math. J., 1957 (9), p. 119.

[10] A. Grothendieck, "Sur une note de Mattuck-Tate," Crelle, 1958 (20), p. 208.

[11] J. I. Igusa, "On some problems in abstract algebraic geometry," Proc. Nat. Acad. Sci. USA, 1955 (41), p. 964.

[12] S. Kleiman, "A numerical theory of ampleness," thesis, Harvard, 1965, to appear in Annals of Math.

[13] S. Kleiman, "A note on the Nakai-Moisezon test for ampleness of a divisor," Ann. J. Math., 1965 (87), p. 221.

[14] K. Kodaira, "A differential-geometric method in the theory of analytic stacks," Proc. Nat. Acad. Sci. USA, 1953 (39), p. 1268.

[15] K. Kodaira and D. C. Spencer, "A theorem of completeness of characteristic systems of complete continuous systems," Am. J. Math., 1959 (81), p. 477.

[16] S. Lang, <u>Abelian Varieties</u>, Interscience-Wiley, N. Y., 1959.

[17] S. Lang and A. Neron, "Rational points of abelian varieties over function fields," Am. J. Math., 1959 (81), p. 95.

[18] T. Matsusaka, "Theory of Q-varieties," Publ. of Math., Soc. of Japan, No. 8, Tokyo, 1965.

[19] B. Moisezon, "The criterion of projectivity of complete algebraic abstract varieties," Doklady Akad. Nauk, Math. Series (28), p. 179.

[20] D. Mumford, Geometric Invariant Theory, Springer-Verlag, Heidelberg-Berlin-N. Y., 1965.

[21] J. P. Murre, "Contravariant functors from preschemes to abelian groups," Publ. Inst. Hautes Et. Sci., No. 23, Paris, 1964.

[22] M. Nagata, Local Rings, Interscience-Wiley, N. Y., 1962.

[23] Y. Nakai, "A criterion of an ample sheaf on a projective scheme," Am. J. Math., 1963 (85), p. 14.

[24] Y. Nakai, "On the characteristic linear systems of algebraic families," Ill. J. Math., 1957 (1), p. 552.

[25] H. Poincaré, "Sur les courbes tracées sur les surfaces algébrique," Ann. École Norm. Sup., 1910 (27).

[26] GAGA: J.-P. Serre, "Géométrie algébrique et géométrie analytique," Annales Inst. Fourier, 1955 (6), p. 1.

[27] J.-P. Serre, "Faisceaux algébriques coherents," Annals of Math., 1955 (61), p. 197.

[28] J.-P. Serre, Groupes algébriques et corps de classes, Hermann, Paris, 1959.

[29] J.-P. Serre, Corps locaux, Hermann, Paris, 1962.

[30] J.-P. Serre, "Sur la topologie des variétés algébrique en caractér istique p," Symp. of Alg. Top., Mexico, 1956.

[31] J. Tate, "Rigid analytic spaces," mimeographed notes secretly printed by Inst. Hautes Ét. Sci., Paris.

[32] O. Zariski, Algebraic Surfaces, Springer-Verlag, Heidelberg-Berlin, 1934.

[33] O. Zariski, and P. Samuel, Commutative Algebra, Van Nostrand, Princeton, 1958.

[34] A. Mattuck and J. Tate, "On the inequality of Castelnuovo-Severi," Hamb. Abh., 1958 (22), p. 295.

Milton Keynes UK
Ingram Content Group UK Ltd.
UKHW040718260124
436738UK00001B/63

9 780691 079936